HIGHLAND BRIDGES

Kylesku Bridge, Sutherland.
Courtesy of Highlands and Islands Development Board

"HIGHLAND BRIDGES"

Gillian Nelson

Bridges should be part of the way, the way continued over
water, convenient, beautiful and able to withstand the years.

Palladio

ABERDEEN UNIVERSITY PRESS
Member of Maxwell Macmillan Pergamon Publishing Corporation

First Published 1990
Aberdeen University Press

© Gillian Nelson 1990

British Library Cataloguing in Publication Data

Nelson, Gillian, 1932–
    Highland bridges
    1. Scotland. Highland region. Bridges
    I. Title
    624.2094115

ISBN 0 08 037744 0

PRINTED IN GREAT BRITAIN
THE UNIVERSITY PRESS
ABERDEEN

To TW and RFJ
and all my friends in the Highlands

Contents

Illustrations

Figures

All the diagrams were drawn by Felix Nelson, except for figure 5 which is an illustration from *The Carpenter and Joiner's Assistant* by J Newlands, published in 1860.

Maps

The maps were prepared by R D Nelson.

Acknowledgements

In preparing this book I have received many small items of information from strangers who spoke to me when I was looking at bridges. I cannot thank them individually, not knowing their names, but I hope that, if they read this, they will take it as acknowledgement of their kind help and friendly interest.

My particular gratitude is due to two civil engineers, who have helped me with practical advice, instruction and encouragement—Neil McInnes who works in Inverness, and Roland Paxton who is Scottish Group Convener of the ICE Panel for Historical Engineering Works.

I am also grateful to the following for the active interest they have taken in my project: Mr Allan of the Caithness Roads Department, Elizabeth Beaton of Hopeman, A D Cameron, M Chrimes, librarian of the ICE, Ronald Curtis, Mr Danby of the HRC, J W Dymock of the NTS, Robin Fleming, Stuart Forbes, Col Alan Gilmour OBE, Ian Hay of Ross and Cromarty Roads Department, Miss Houston of Huna Mill, John Hume of the Historical Buildings and Monuments Commission, Dr Iredale, Moray District Archivist, Mr Irwin of West Ross Roads Dept, John Kerr of Old Struan, R Lee of Grampian District Council, Murdo MacDonald, the Argyll and Bute Archivist, Mrs R McGregor of Inverary, Lady Mary McGrigor of Dalmally, Donald Mack, B MacKay of the SDD, Donald MacKechnie OBE, David MacKenzie of HRC, Mr Macneil of SDD, Charlie Menzies of Glen Cassley, Judith Robertson of HRC, Stephen Rosenberg, the RCAHMS staff in Edinburgh, R J Spoors of Scotrail, Mr Strong of Lochaber Roads Department, Ian Sutherland of the Wick Heritage Centre, Mr O T Williams of Sir O Williams and Partners Ltd.

Finally, I owe a great deal to the help and time unstintingly given by my son, Felix, who drew the diagrams and my husband, David, who prepared the maps.

GN

Introduction

The Building of Bridges

Bridges have existed since man first needed to cross a river he could not wade through or swim. Nature gave him hints: a fallen tree over a stream; strands of creeper looped across a ravine; and, in a few places, often by the sea, natural stone arches that had formed through the weathering of rock. Using these hints man applied his ingenuity.

There are essentially three forms a bridge can take, and for all three you must first provide strong supports and anchorages on the river banks. These are called abutments.

Type One: beam bridges On abutments set into firm ground you place a beam of wood or metal to bridge the gap. (Originally the word beam meant tree, as in whitebeam.) If the river is too wide for one beam to stretch right over, you have to use several and to support them, where they join, on pillars set into the river bed. These pillars are called piers, and the stretch of deck between abutment and pier, or between one pier and the next, is a span.

In this type of bridge the weight is applied by the beam downwards on to the abutments and piers (Figure 1). Some primitive bridges can still be seen with low, dry-stone piers and single flagstones providing the beams. A one span bridge of this sort is a clam, one with several spans is a clapper. A fine example is the Torridon clapper, see page 184.

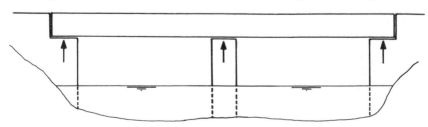

Fig 1 Beam bridge with the weight downwards on to abutments and pier

xiv

Type Two: an arch bridge   A simple example is an arch of stones built between river banks in such a way that the spring of the arch—that is the lowest point of the curve—is set into the abutment. For all their appearance of airy grace, arches are strong structures. The downward weight of the walls above and the traffic crossing is converted into a sideways push that runs round the archring (Figure 2). In order for the arch to become overloaded the forces would have to be large enough to crush the stone sections of the arch or push away the abutments. The stones in a ring on the outer face are called the voussoirs and the continuation underneath, making the vault, is the soffit.

The strength of the arch is well seen in some ruined bridges where only the ring remains and yet can be walked over. For instance, the famous Carr Bridge, see page 71. The walls built over the arch ring and out to the abutments are called spandrels. They are packed with rubble and it is they that give way under lateral pressure. For this reason some dickey old bridges are stayed, that is held together with iron bars, which spoils their looks but prolongs their lives (Figure 3). An example is Bridge of Avon at Dalnashaugh, see page 77. Various techniques have been devised to lessen the spandrels' weight. The Romans built wine jars into them; Telford made them hollow with strengthening inner struts; decorative holes, called oculi, can be pierced through.

There are three common arch shapes. Early builders, notably the Romans, used semi-circles. A reflection in water of such an arch makes a perfect circle. Later, segmental arches became more common. These only use a section of half a circle. There are also elliptical arches. Their

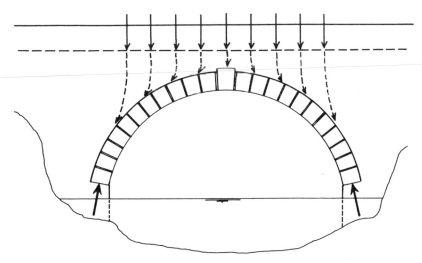

Fig 2   Arch bridge with the pressure taken round the ring

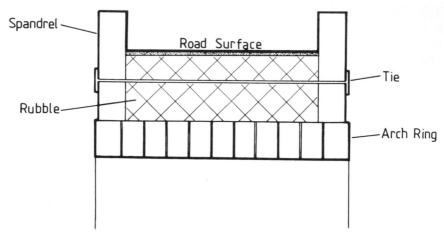

Fig 3   Tie bar inserted above an arch through the spandrels to strengthen them

reflection makes an egg shape, rather than a circle (Figure 4). To build any stone arch you need a wooden frame to support the stones while they are being assembled, for there is no strength in the voussoir ring and soffits until they are complete. The timber frame is known as centering, or false work (Figure 5).

Type Three: suspension bridges   The third option for a bridge builder is to raise the abutments into towers and from these loop across cables from which the road can be hung. The cables have to be securely anchored into the ground behind the towers. The pressure here is down on the abutments, while the cables are in tension (Figure 6). The deck is free to move and, in simple bridges of this type, it does. When used for modern traffic, ways have to be found of making the deck rigid, unable either to sway or to twist. We are inclined to think of suspension bridges as modern, but the Chinese were building them with chains as long ago as the first century AD.

The early history of bridge building is lost to us, but we know that sophisticated bridges were being built many thousands of years ago. At Diz in Khazistan there is a bridge with twenty arches which may date from 350 BC. Xerxes' famous bridge over the Dardanelles of 480 BC was designed by the first civil engineer known by name to us, Harpales. He anchored 360 ships in two parallel lines and linked them with cables of papyrus and flax tightened by capstans on the shores. The Greeks, though great builders, did not much use the arch. What they did build were corbelled, that is built out from each side, stone overlapping stone until they met in the centre (Figure 7). Such arches tend to look cramped

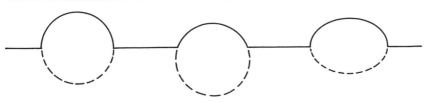

Fig 4   Three arch shapes: semicircular, segmental and elliptical

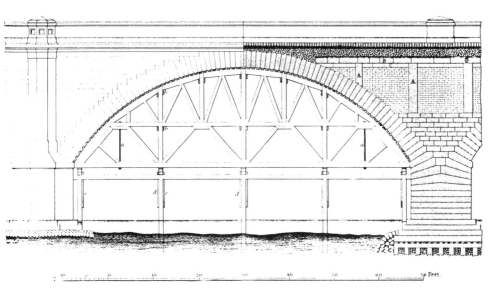

Fig 5   Drawing of the centering for Telford's bridge over the River Don in Aberdeen. The centering was designed by the contractor John Gibb & Son of Aberdeen. From *The Carpenter and Joiner's Assistant*, J Newlands (1860). *Courtesy of R Paxton*

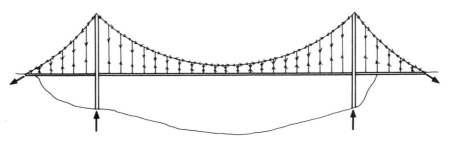

Fig 6   Suspension bridge, with cables in tension

and the method does not allow large spans. The Romans were colonisers and needed roads. Their bridges were built to last and many have, down to the present time. One of the finest is Pont du Gard at Nîmes, dated about 19 BC. It has three tiers of arches, six in the bottom row, eleven in the second and thirty-five in the top carrying water. Only the top tier uses mortar. All the arches are semicircular. Egyptians and others had used a gypsum mortar but the Romans developed the idea of hardening mortar by incorporating volcanic dust and burned lime thus making a strong cement called pozzolana. This cement has the marvellous quality of setting in water, so was ideal for bridges.

The medieval bridge was often a water platform for all sorts of buildings from chapels to shops and prisons. The piers might be only piles of loose stones built up on the easiest places in the river. This resulted in uneven spans, a noticeable feature of such bridges, and, because the piers had to be thick, the amount of navigable river was much reduced. They were in fact the opposite of our century's spare, functional bridges.

With the dawn of a new interest in science in the seventeenth and eighteenth centuries, the theory of bridge building developed again. The segmental arch began to take over. Masonry work became finer, detailing more precise and city fathers vied in spanning their rivers handsomely.

Galileo in 1658 wrote a book on stress in structures, and Robert Hooke, in 1678, started a series of experiments involving compression and tension and produced a formula that was the basis of all subsequent work on stresses. Palladio, who has given his name to a style of classical architecture, designed four trusses. He pointed out that they were economical, could span great lengths, and utilised short beams.

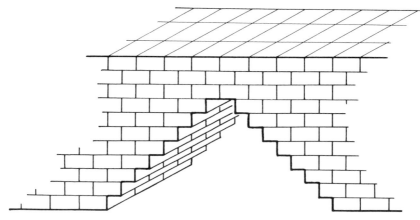

Fig 7   Corbelled arch

A truss is a sophisticated form of beam. It is based on a useful property of triangles. Whereas you can change the shape of a square into a parallelogram without altering the length of its sides, a triangle you cannot (Figure 8). This geometrical property is used to build light strong beams (of wood or metal) which will not deform excessively under pressure. The Americans were great pioneers in this field. The simplest and best known truss is the Pratt. Most are of metal, the early ones of cast iron and the later ones of wrought iron and steel.

In the middle of the nineteenth century it began to be feasible to build iron bridges because of improvements in the smelting and casting methods. Iron has an immense advantage over stone—its comparatively light weight. This was a particular boon to railway engineers because their bridges often had to span large distances. This is due to the fact that trains must run on as level rails as possible and cannot deal with steep gradients. You cannot have a sloping approach to a railway bridge.

The development of wrought iron and steel, which withstand tensile forces better than cast iron, allowed for the re-emergence of suspension bridges. The first one in Britain is reputed to be the Winch Bridge over the Tees near Middleton, built in 1741 as a footbridge for miners. It did not have a stiffened deck and one traveller described 'the tremulous motion of the chain' and seeing himself 'suspended over a roaring gulph on an agitated, restless gangway'. The problem of the restless gangway is the main difficulty with suspension bridges. The simplest have flexible decks, see, for instance, the bridge at Bonawe, page 198. For wheeled traffic this will not do. The deck has to be stiff iron and yet not put too much weight on the cables. The Forth Road Bridge, over which you may have come driving north, is one of several modern suspension bridges that solve this problem by having a light steel deck on transverse lattice girders.

The older Forth Railway Bridge, built in 1890 and designed by Benjamin Baker and John Fowler, uses an entirely different technique, that of cantilevers. This method is a variant of Type One, the beam. Just as the branch of a tree is held by the trunk and will support a given weight, so in a cantilever the weight is carried on arms extending from the abutments and unsupported by piers. Often these arms in turn

Fig 8   Differing properties of squares and triangles

support central beams, as in many motorway bridges (Figure 9). With the Forth Bridge, the distance to be spanned demanded six cantilever arms, projecting from three iron towers, and themselves carrying truss beams. The theory is illustrated in the photograph of an experiment set up by Baker. A good example of a cantilever bridge in the Highlands is Connel, see page 36.

There is yet another form of iron bridge. This is the bow string truss, in which the road is suspended off the arch which it ties (Figure 10). A magnificent bow string bridge is the railway bridge at Speymouth, see page 91. There are also, in this region, three unusual concrete bow string bridges, one on Rannoch Moor, see page 196.

In the twentieth century we enter the era of concrete, although as early as 1756 John Smeaton, asked to build the Eddystone Lighthouse, experimented with rock dust from the Rhine similar to that from Pozzoli that the Romans used. They used cement to stick stones together, but modern cements are cohesive rather than adhesive.

Because of concrete's low resistance to stretching, attempts were made to strengthen it with wrought iron bars inside. At the Paris Exhibition of 1854 a boat with a reinforced concrete hull was shown, but it was a gardener, called Joseph Monier, who popularised the material by selling flower pots made of it. Thaddeus Hyatt in America, and François Hennebique in France made scientific experiments with reinforced concrete, and Hennebique was responsible for one hundred bridges, the most famous being the Risorgimento Bridge in Rome.

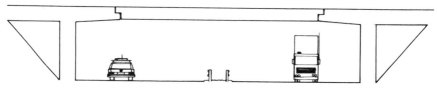

Fig 9   Bridge with two cantilever arms supporting a central beam

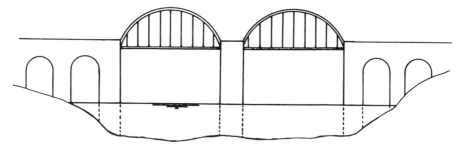

Fig 10   Bow string arches in which the deck ties the arch from which it is suspended

1   Baker's demonstration of the forces in a cantilever bridge.
Courtesy R Paxton

Another master was a Swiss named Robert Maillart, and possibly his most elegant design is the Schwandbach Bridge in Switzerland—a virtuoso performance by any standard.

One problem with reinforced concrete is cracks which are caused by differing stress performances of the rods and the concrete. This difficulty was solved by Eugene Freyssinet (1879–1962) who subjected the concrete to initial compression to neutralise the tension it would be subjected to later when it took a load. He did this by stretching the reinforcing bars and then relaxing them once set in the concrete, so that the concrete was compressed and kept that stress throughout its life. This is known as pre-stressed concrete.

Bridges built from this material can have spare proportions with low elliptical arches, and slender piers—a far cry from the Romans' monumental stonework. Such bridges are, in effect, both iron and stone. The elegance of their stone curves is directly attributable to their steel stiffening. The most remarkable of such bridges in this area is Kylesku, see frontispiece.

Such is a bare outline guide, written by a layman, to the history and science of bridge building. To those interested in reading deeper, I can recommend in particular H J Hopkins' *A Span of Bridges* and J E Gordon's *Structures*, and other books listed in the bibliography.

The universal appeal of bridges is due to the fact that they are expressions of man's constructive imagination, and of his wish to cross divides and unite opposites. The almost mystical importance of bridges is implied in the title given to Roman rulers—pontifex maximus, the greatest bridge-builder. The early lakeside dwellers of North Italy saw the bridge as a symbol of the link between men and gods. Today pontifex maximus is a title still used by the Pope.

A bridge has the useful function of shortening distances and so speeding transport and helping commerce, and also the less tangible property of bringing people together. It is democratic as opposed to a drawbridge which is divisive.

Bridges are both humble, serviceable things—a way over water for all to use—and at the same time symbols of man's conquest of natural forces. The arch, held in place by its own counter-thrusting forces, expresses both movement and stability. This is evident in all great arched buildings—from Roman temples to nineteenth century train sheds—but in a bridge, which is all arch, with no roof, walls or doors to obscure its form, the effect is all the more compelling.

You may prefer the simplest bridge of all—the clam, the carefully chosen flat stones placed over a burn in the hills and so bedded down in the turf of the banks as to seem almost as permanent as the rocks around, trodden over gratefully by generations of shepherds and sheep. Or you may like best the functional bridges of our own time, such as that over the Beauly Firth seen as it fades against storm clouds and night, its warning lights like red stars. Perhaps you take most pleasure in the friendly simplicity of a stone arch, or in ornate, cast iron lattices, or the long legs of a viaduct spacing out a valley, or . . .

Whatever your choice is, I wish you good looking.

2   Clam bridge in Glen Affric

Highland Roads

The very name of high lands suggest that this area is not ideally suited to roads, and until the mid 1700s there were few.

Such stony causeways and bridges as did exist had to be kept in repair. Money for bridge maintenance was raised by pontage. A typical charge in the 1600s was forty pence for a millstone to cross, and eight pence for each cart of wine or timber—clearly the sort of loads that did the equivalent damage of our juggernauts. The actual labour was the responsibility of householders who were warned when it was needed by an announcement in church. They had to provide the tools, carts and horses and to work six days a year for three years, and four days a year thereafter. Defaulters could be fined, or 'punished in their persons'. This was made law by an Act of 1669, but the very next year it was modified. No road work could be demanded at seedtime or harvest, and those who lived at a distance from the road to be repaired could commute their duty into money to hire labour. The payment set was six shillings for a man and twelve for a horse. Few ordinary people would have had such a sum to spare and remote, exposed stretches of the track may well have gone by the board. As far as the standard of repairs goes, the same act specifies that the roads be 20 feet wide and kept at a level that horses and carts could travel on them all the year, which in such a climate as ours is a high standard indeed and may well not have been generally maintained. This is hinted at by the 1670 Act also prohibiting the 'plowing up or laying stones, rubbish or dung thereupon', or 'pooling the same, or turning in or damming water thereupon'. If such frightful acts occurred, they were to be deemed as having been done by 'the labourers of the land next adjacent', which seems a fair bet. For the realigning of roads or making of needful new bridges the JPs could levy a tax not exceeding ten shillings Scots on each £100 of rentable value each year. In other words, the rich landowners paid.

In all probability, the country would have been a mixture: regions with roads passable all year, others with tracks that were all right in dry periods, and places where one followed the contours of the hills and animal paths as well as one could. Very likely the route of a given track varied depending on the time of year, avoiding marshes in wet weather, keeping low in snowy times, and so on. On the rugged western coasts almost all transport was by sea, and the settlements hugged the shore-line.

Long distance tracks were imperative for the driving of cattle in summer to the beef markets in the Lowlands. The cattle swam the sea inlets and the rivers. The drove roads criss-crossed the country, many going to considerable heights, but they were not built in any sense and were no doubt difficult to use in certain weathers and places. If you walk them now you will pass the ruins of many small villages, or townships, formerly the homes of thousands of farming people. Obviously such communities must have had local tracks for the passage of horses, farm animals and carts; although in some of the craggier areas wheel-less carts were used. These were a sort of sledge with a housing on top and were pulled by men or animals. As late as 1799 it was reported there was hardly a wheeled cart in Caithness.

It is General George Wade who is remembered as the maker of Highland roads. He was an Irishman of a family settled in that country by Cromwell. He had fought in Spain and became MP for Hindon in Wiltshire, and later for Bath. He apparently made only one memorable speech in the House, that was in 1733 on the subject of Courts Martial.

He certainly began the process of providing our modern road system as the servant of a Government wishing to have better control of potentially rebellious highlanders. Like the roads going north from Rome, Wade's were the fingers of a colonising hand resting in London. He started with the Fort William to Inverness road in 1725 and approached Inverness through Dores after great exploits in blasting the rocks. By 1734 he had built 250 miles of roads and forty bridges. Wade then left the Highlands. What he had built was a framework for moving armies north and south quickly. He did not make the network of roads that legend ascribes to him.

His successor was Major Edward Caulfeild, who was also an Irishman, and served in Harrison's Regiment. He worked on the military roads until 1767 and extended them to the west and to the east, and linked the prestigious new Fort George at Ardersier into the network. Caulfeild probably wrote the couplet about Wade:

> If you'd seen these roads before they were made,
> You'd lift up your hands and bless General Wade.

Caulfeild left 1,180 miles of road and 682 bridges for his successor, Colonel Skene, to keep in repair and this took most of his allocation of funds. Most of the military roads were to the east and south of the Great Glen, except for the Bernera barracks road (meant to menace Skye from where invasions might come), and the Argyll road to Bonawe.

Some private citizens did some road building, but only on a small scale. The overall lack of roads, even after Wade and Caufeild's labours, is shown by the number of ferries. If you took a ferry from Invergordon to the Black Isle and then another to Fort George, it saved you a day's journey. Now the distance could be covered easily in an hour by road. In 1799 it was not possible to travel north to Inverness without a ferry crossing either of the Tay at Dunkeld or of the Spey at Fochabers. In 1771 John Kirk surveyed the 'king's high road to Caithness' skirting the coast, which argues there was something to survey. On the other hand in 1789 only one carriage tax (the equivalent of our road tax) was taken out by a Mrs Sutherland of Uppat. It cost her £3. 10s. 0d., which was presumably Caithness County's road department income for the year. As one minister commented, 'Roads are made by nature.'

The whole country was hamstrung by lack of good communications at the same time as it suffered the iniquitous clearances. Emigrations from the Highlands reached a peak in the 1760s, when 10,000 left from the Hebrides and Inverness-shire, and there was a second peak at the turn of the century. Long after most people of goodwill had become troubled at the situation, the Government asked Thomas Telford to send them a report on the Highland roads.

Born in 1757, the son of a shepherd who died when he was a baby, Telford is the archetypal country lad made good. He was brought up by his impoverished mother in one room of a two-roomed cottage, sent to the village school, apprenticed to a stone mason, repaired a bridge swept away in a storm, went to London and worked on the stonework of Somerset House, then to Shrewsbury where he was taken up by Sir William Pulteney and started on large scale engineering work. He ended his life as the first President of the Institute of Civil Engineers.

To prepare his report he travelled tirelessly on horseback throughout the summers of 1801 and 1802 and he recommended the building of roads, bridges and harbours to improve the life of the Highlanders who were, by their lack of access to markets, painfully, and in some cases degradingly, cut off from services and ideas that the rest of Britain took for granted. A Parliamentary Commission was set up in 1803 to put into effect Telford's proposals, and he himself was asked to undertake the enormous labour, on top of engineering schemes he had in hand in England, in the south of Scotland and even in Sweden.

He began his great work at the age of forty-six and was assisted by John Rickman, a civil servant in London who acted as secretary and linkman, and by John Hope, a lawyer in Edinburgh. The three worked with amazing and continuing zeal for many years to obtain the necessary land from unwilling landowners, to set up the huge logistical infra-structure to support the construction of so many large building works in a country where you were building the very roads you needed to convey your materials. No compensation was paid for land on the grounds that the landowners would benefit. In some cases they co-operated handsomely, providing stone, timber and labour. In others they procrastinated and went to law. The roads were financed on a half-and-half basis between the Government and local individuals or town councils. Hence they came to be known as Parliamentary Commission roads, or simply Parliamentary roads.

Telford, unlike Wade, considered the needs of the inhabitants and took his roads the way they wanted them to go. Southey wrote, 'These roads have given life to the country,' and one of Telford's deputies called them 'irrigating streams'. On his tour of 1819, Southey saw on the Wick road traders with carts of Worcester chinaware which they had previously carried on their backs; and he saw woolpacks waiting to be shipped at Bonar.

At the limit of extent of the Parliamentary roads, the Statute Labour roads began, that is roads built by men who owed statutory labour to the Council; but as early as 1821 in Sutherland this labour was converted into money to employ surveyors and road makers under the Trustees. The Rosehall to Assynt road was built thus in 1822, and by 1860 Sutherland had 510 miles of road, Caithness 231, and Ross and Cromarty 776. The respective rates for these three counties were £1. 4s. 0d. in the pound, seven pence and nine pence. The rates were based on the population rates per mile which were 50 in Sutherland, 178 in Caithness, and 105 in Ross and Cromarty.

After this immense effort in the first part of the nineteenth century to provide a road system for the Highlands, nothing more was done on a substantial scale until the arrival of the motor car coincided with the belated realisation in London that the highlands were again suffering the disastrous economic and human effects of poor communications. In the 1920s and 1930s the A9 was driven north and many smaller roads were also made or re-made. The work was directed by Major Robert Bruce, the chief engineer. It is instructive to see, in old photographs taken at the time, the sort of minor bridges that Bruce's team replaced. Many were pleasant, solid stone arches of Parliamentary design; but there were also wooden trestles such as at Frithe, near Tomatin, and metal trough structures as at Balavil, north of Kingussie, and even simpler timber bridges of a type now used exclusively for pedestrians, such as at Coachan and Chuirn on the approach to the Drumochter Pass.

At this period among the many roads repaired and partially re-aligned was the one on the west side of Loch Ness which gave rise, in 1933, to the re-awakening of the monster to greet the now more mobile tourist.

Most of these pre-war roads were single track, with passing places marked by metal diamonds painted white, on poles. In the remoter areas these roads are still in use, though, for the most part, slightly improved at the blinder bends and steepest hills which used to be tricky, to say the least.

Even in the 1970s the A9 itself was a fairly narrow main road, as can be seen where sections have been left beside the new road as laybys. To be caught behind a coach and two caravans leading a queue of cars, as our family car, holiday-bound, laden with tents and children, once was on a never-to-be-forgotten, hot, and interminable journey, could mean 30 miles of crawl punctuated both by 'fast johnnies' (as my father called them) who whizzed up the fuming line of cars in a terrifying way, and by the weary, repetitive question 'Are we nearly there now, Mummy?'

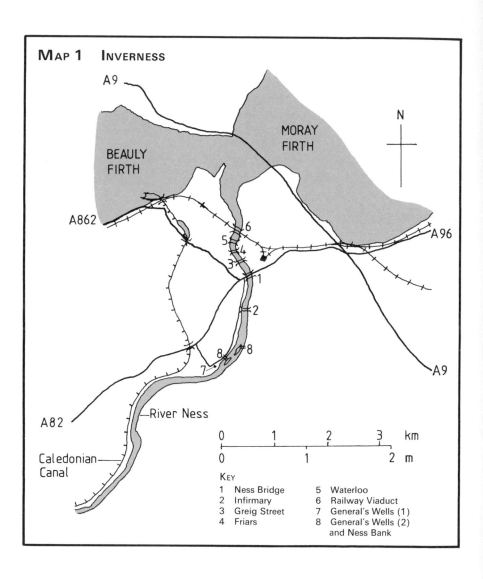

MAP 1   INVERNESS

A9

MORAY
FIRTH

BEAULY
FIRTH

A862

A96

6

5
4
3
1

2

8
8
7

A9

River Ness

A82

Caledonian
Canal

| 0 | 1 | 2 | 3 | km |
| 0 | | 1 | | 2 m |

KEY
1 Ness Bridge   5 Waterloo
2 Infirmary   6 Railway Viaduct
3 Greig Street   7 General's Wells (1)
4 Friars   8 General's Wells (2)
                       and Ness Bank

I

Inverness

Map 1

NESS ISLANDS BRIDGES (1)	1853	James Dredge	*NH 658 434*
RAILWAY VIADUCT	1862	Joseph Mitchell	*NH 663 460*
INFIRMARY BRIDGE	1879	C R Manners	*NH 664 445*
GREIG STREET BRIDGE	1881	C R Manners	*NH 664 454*
WATERLOO BRIDGE	1895	MacKenzie and Murdoch Paterson	*NH 662 459*
NESS BRIDGE	1961	Murdoch MacDonald	*NH 665 451*
FRIARS BRIDGE	1987	Jamieson, MacKay & Ptnrs	*NH 662 456*
NESS ISLANDS BRIDGES (2)	1988	Dinardo & Ptnrs	*NH 661 436* and *NH 664 439*

The Highland city of Inverness is ancient and in a strategic position, being on the coast, at the head of the Great Glen, and at the entrance to the Beauly Firth. Alexander II of Scotland built the first bridge here before 1249, approximately at the place in the centre of town where the Ness Bridge now stands. There followed a succession of timber bridges and in 1684 the first stone bridge which had seven arches. Judging by old prints, it was a graceful structure. It was very much part of town life with people coming for water and the laundry women working on the banks and bleaching the linen on a green where the cathedral now stands. The pontage, or toll, on the bridge was a bodle (1/6 of an old Scots penny) for a man and a penny for a horse. To avoid paying many people waded across, and some earned money carrying others over. A macabre feature of this old bridge was a room in one of the piers down at water level which was used first as a prison cell and later as a place to shut up poor lunatics. This bridge came down in January 1849 during a violent storm and after a night in which the helpless townspeople had watched the supports loosening. Its lights burned to the moment it fell

1

just after six o'clock. A drunken sailor—rolling home quite oblivious of the state of affairs and watched in horror by the crowd—was the last person to cross it.

There was considerable controversy over the building of a replacement. An engineer called Joseph Mitchell hoped he would get the contract. He was well known in the area as a road and bridge builder. His father, John, had been Telford's deputy for many years and Joseph had inherited his job and his skills. He was a vigorous, amusing, friendly character as his book of reminiscences shows, and an excellent engineer. During the three days of torrential rain that preceded the collapse of the old bridge, he had been active in helping to repair breaches in Telford's canal nearby. Afterwards he put up a temporary timber footbridge. To Mitchell's chagrin the contract went to a Mr J M Rendel, a younger man whom Mitchell had helped in the past. Rendel, asserting it was unsafe to build piers in the Ness because of its tendency to violent floods and its tidal scour, designed a suspension bridge. He estimated it would cost £15,000 but, in the event, and after five years' work, it cost double that sum, partly due to the workmen striking for a three shilling a week rise. They then earned one guinea. Mitchell said he could have built a bridge over the Ness with piers for much less, and in fact did so in 1862.

Rendel's bridge lasted until 1961 when it was removed because it was too narrow. The present bridge was designed by Murdoch MacDonald and Partners, with Duncan Logan Ltd as contractors. It has three spans of 62 feet 6 inches, 120 feet, 62 feet 6 inches and is built of pre-stressed, reinforced concrete. Its overall length is 253 feet. The central span is two 30 feet cantilevers, supported on piers, and a suspended 60 feet span in the middle. Various facing stones have been used—Shapgranite, Corrennie granite from Aberdeenshire, and Tarradale sandstone. Twelve thousand tons of concrete were used, and 280 tons of steel. It cost £250,000 which was an eight-fold increase on Rendel's price. The decorative detailing of this bridge includes the part of the pylons from the suspension bridge which supported the roller bearings.

This central Ness bridge is flanked on both sides by graceful cast iron suspension bridges, painted pale grey, and designed by C R Manners. The upstream Infirmary Bridge was built first with a span of 200 feet and two shorebays of 42 feet. It was judged by some to be too frail. Greig Street bridge, built in 1881, is similar but more substantial. It cost £1,400 but was not built without litigation over fishing rights. This is the bridge that many visitors to Inverness remember best. It makes a harmonious focal point for a riverside walk, either framing the castle or Ben Wyvis, its fluid lines contrasting with the extraordinarily many and varied church spires and towers along the Ness banks. Its looks depend

3   Greig Street Bridge

chiefly on the tapered lattice pylons linked by a decorative metal arch, the triangular truss, and the gentle but decided arch of the deck. The pylons are on pairs of cast iron pillars (now tied with metal bands), and the cables are anchored in the abutment walls. You will notice new stone work here and the ramp. These are renovations recently carried out. The deck is wooden slats set on transverse girders into which the suspension rods are fixed with turnbuckles and eyelets, and which also hold the curved stabilisers. In the central section these girders extend beyond the deck and there the rods are screwed directly into them. This detail is different from the Infirmary Bridge and may have been a modification added to strengthen this second bridge. The iron work was cast at the Rose Street Foundry.

In 1881 the Foundry was still known by its earlier name of Northern Agricultural Implement and Foundry Company; it has now become A1 Welders. A great deal of the iron work needed in the Highland region was made at Rose Street. Its impressive buildings in Academy Street and Rose Street were designed in 1893 by Alexander Ross who also designed the cathedral. They are decorated with mosaic panels showing industrial scenes.

Downstream of Greig Street the riverside road on the right bank passes a terrace of small houses and reaches Friars Bridge. Its name is due to the Dominican Priory which existed near here from 1233. A pillar from the Priory remains near the Telephone Exchange in a small graveyard. This bend on the river was used for salmon fishing from the bank. The nets cast out are said to be shot—so the place is called Friars Shott. The three span pre-stressed concrete bridge was designed by Jamieson, MacKay and Partners with Edward Nuttall Ltd as the contractors. It is 146 yards long with spans of 41, 64, 41 yards. The building of this, and all the other modern Ness bridges, involved the construction of coffer dams to exclude the sometimes powerfully racing tidal water while preparing the bases for the piers and abutments. For Friars Bridge 180 steel piles were driven 33 yards into the riverbed.

If you walk under the bridge on the path provided, you will see that it is constructed from two separate arches, each with a deep central girder and wide upper flanges (or a trough with a raised rim) and set on oval piers. The abutments are faced with vertically grooved panels. Otherwise there is little decoration and the curving rise and fall of this low, smooth bridge is left to speak for itself. The railings are painted an unobtrusive brown and the lamp-posts are functional, fitting and pleasant—unlike many bridges where the lighting arrangements are out of keeping with the rest.

This is painfully the case with the next bridge down the river. This iron truss bridge, of double Warren type, is called Waterloo Bridge, and was built in 1896 by John MacKenzie and Murdoch Paterson to replace an earlier wooden bridge called Black Bridge (1808). It is 118 yards long and has five spans on pairs of cast iron columns. These were extended up above the sides to make supports for the gas lamps, but now unsympathetic electric heads have been stuck on. The iron work was

4   The underside of Friars Bridge

again cast at Rose Street. The stone abutments are finely made of dressed stone with battered wing walls, a rounded coping stone and short square pillars to finish. Nowadays it looks a rather weary old black bridge and after the floods of February 1989 when the water rose 11 inches above the previous record height, it had to be strengthened. When newly built and freshly painted with its gas lamps reflected in the river, it must have been attractive.

Close by you will see a railway bridge being built to replace the fine stone one swept away in Febraury 1989. Built in 1862 by Joseph Mitchell for the Inverness and Ross-shire Railway, it had five segmental spans of 75 feet each over the water (and these fell), and a further six spans on the banks, which remain—a total length of 669 feet. Mitchell was particularly pleased with this bridge because, confuting Mr Rendel, he had supported it on piers and because it only cost £13,500. In fact, he called it a 'source of pride and satisfaction'. It is somewhat ironic that his stone arches did eventually fall, albeit after 127 years of strenuous life pounded by trains, by flood and by storm, and fell after just such a three day deluge as caused the destruction of the old Ness Bridge, whereas Rendel's suspension bridge did not collapse but had to be removed due to the demands of modern traffic.

To find the oldest Inverness bridge still existing, you must return up river past Infirmary Bridge, the salmon fishing station, and a pedestrian suspension bridge to the islands. Beyond this, at a sharp right bend in

NESS VIADUCT

Five Arches of 75 Feet Span
and Four Land Arches of 20 Feet Span

5   Mitchell's Viaduct photographed when newly built, with the old Black Bridge behind. Courtesy of ICE

6   A bridge no more: the viaduct in the River Ness, February 1989

7   The old General's Well Bridge, re-erected in the park

the road, there is a park on the left which has a miniature railway for children to ride. This little train crosses an exceedingly pretty white suspension bridge over a dry gully. Until recently this bridge spanned the Ness where the modern suspension bridge now is, and it had a twin on the far bank. They were called General's Well Bridge, and Island Bank Road Bridge.

The two were designed by James Dredge in 1853. They replaced earlier bridges which were put up in 1829 so that the islands could (as was reported in the local newspaper) 'be laid out in graceful and varied walks' which would 'form a lasting ornament to the town, a powerful attraction to strangers, and a source of healthful recreation and enjoyment to the inhabitants'. One imagines the crinolines and top hats, followed perhaps by children bowling hoops, taking healthful recreation in the dappled shade, with one eye on the tourists the new bridges might draw. These bridges fell in 1849.

Dredge worked at a period when many designers were experimenting with suspension bridges. You will notice that on this small bridge the rods are diagonal to the pylons, and that the iron ribbon of cables lessens in width as each rod is linked into the deck. Starting with 6 it reduces to 5–4–3–2. In the centre there is a short double span. In effect the cable and rods are supporting the cantilevered deck and, as there is successively less weight to carry, the tension member can be reduced in size. None of the measurements on this bridge are very uniform. The curved lateral girders under the deck are not equidistantly spaced. The

8   Detail of cable and rods on General's Well Bridge

overall length when *in situ* over the Ness was 97 feet 3 inches. Its twin (now scrapped), the Island Bank Road Bridge, was slightly shorter at 83 feet. The slatted deck is 70 inches and the railing 39 inches tall.

A much larger version of this bridge is the now partly ruinous bridge at Aberchalder over the River Oich (see page 16). Dredge's design was never taken up and it is good to see this small historic, and charming, bridge has been conserved. Perhaps more attention could be drawn to it. It is sad that its twin has disappeared and also that no money can be found to paint, preserve and repair the larger Dredge bridge. Until recently, I am told, there were two similar bridges by Dredge in the Republic of Ireland but one has recently fallen.

The places occupied by the Dredge bridges over the Ness have been taken by strong modern suspension footbridges. They are goodlooking and have been designed to fit the small park-like scene. Messrs Cruden and Tough of Dinardo's designed them. A comparison of the outline of piers, cables and rods of the two types of bridge shows at a glance their different construction. The new bridges have decks of hardwood slats, as the originals had, and details like the neat railing running past the end of the bridge proper and its finish in a steel sphere keep a faintly Victorian air. The hanger rods are 22mm mild steel and the cables 35mm spiral strand. The towers are plain and quite elegant rising to a height of 8 yards. The existing ashlar abutments were retained but strengthened with concrete 'hearts', and the unobtrusive lights on the crossbar are welcome.

Further bridges that should be mentioned in Inverness are Clachnaharry rail bridge, and the canal bridges. Kessock Bridge, not strictly in the town but nevertheless dominating it, is dealt with in the next chapter.

II

Kessock to Fort William

Map 2

As you approach Inverness from the south and come to the crest of the long hill down to the city, you see a mysterious, exciting and enormous expanse of land, water and sky with the sea running inland through the mountains and the mountains receding westwards into the clouds. It must be one of the finest views in Europe. To your right the estuary widens between the Black Isle and Moray coast; to the left steep hills compress Loch Ness into its trough. At the foot and in the centre is the low lying city with the Kessock Bridge arching away over the firth.

This chapter starts at Kessock and takes you down the Great Glen to Fort William at the western end—a route that is both inland and towards the sea.

KESSOCK BRIDGE	1982 Crouch and Hogg Arup and Ptnrs	*NH 665 475* on the A9, tourist park on the north side

This is the only cable stay bridge in Britain. Its design presented considerable problems. It had to withstand winds of up to 99 mph, a tide race of 8 mph in water 13 yards deep, and even possible seismic tremors since Inverness is on a fault line spasmodically active. There had to be a channel for navigation. It was built to a German design and cost more than £33 million. It is 65 yards long and is supported by piers built on steel piles themselves driven 65.6 yards deep into the seabed. There are sixty-four cables of spiral strand steel, each composed of 271 wires. The steel towers they are suspended from rise 43.7 yards above the deck. They are hollow with stairs inside and are topped by navigation lights.

Coming from the south, the town side, the curve, bank and rise of the approach and bridge is smooth and exhilarating. The cables, in a triangular pattern are like feathers on an arrow or rigging on a mast. On the northern approach road you get a more level and less exciting vista.

9

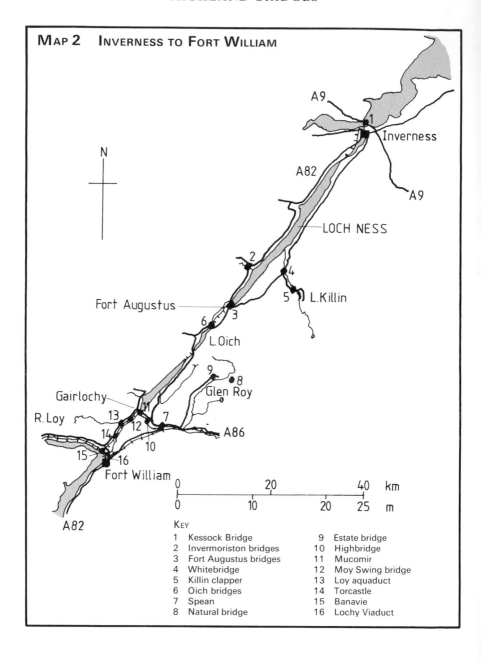

MAP 2   INVERNESS TO FORT WILLIAM

A9

1 Inverness

A82

A9

LOCH NESS

2

4

L.Killin

Fort Augustus

5

6 3

L. Oich

9 8

Glen Roy

Gairlochy

R. Loy

13

11

7

14 12

10

A86

15 16

Fort William

| 0 | | 20 | | 40 | km |
| 0 | 10 | | 20 | 25 | m |

A82

KEY

1	Kessock Bridge	9	Estate bridge
2	Invermoriston bridges	10	Highbridge
3	Fort Augustus bridges	11	Mucomir
4	Whitebridge	12	Moy Swing bridge
5	Killin clapper	13	Loy aquaduct
6	Oich bridges	14	Torcastle
7	Spean	15	Banavie
8	Natural bridge	16	Lochy Viaduct

9   Kessock Bridge from the north side

It is above all a bridge to look out from, for it gives the geographical lie of the land usually only gained from the air. Across the estuary the River Ness enters the sea, and, to the right, is the canal entrance. You can see the clay embankment that had to be constructed in the sea before the lock, then the largest in the world, could be built in what the contractor called 'a morass of mud'.

Below the car park is the village of North Kessock from where you have a different view of the bridge and its detailing is more noticeable—the lamp standards, the steel and rubber fenders and the crash barriers. Beside the hotel is a sandstone cottage and wide barn door. This was the ferryman's house and the carriage shed. There has been a ferry across these narrows since at least the 1400s, and until after 1900 it was a sailing boat. Southey said the Kessock Ferry was the best in Scotland 'but the best ferry is a bad thing'.

It would be hard to find a bridge site that more exactly fitted the classic case. Kessock spans an estuary of fast, difficult currents in a place known for wild weather. It links the main body of the Highland region to the hinterland north and west, particularly as it is partnered by the Cromarty Bridge. Soon there will be another link over the Dornoch Firth.

From Inverness take the A82 south down Loch Ness-side. Where the River Moriston narrows and falls rapidly in a series of low waterfalls are the **INVERMORISTON BRIDGES** (*NH 419 165*). Both are masonry arches and, seen from down stream where there is a Victorian summer house, they make a picturesque view. The older, now partly ruined, was ascribed to Telford who certainly built a road here between 1808 and 1811, but it is now considered to be a mid nineteenth century structure. Telford's spans were 50 and 46 feet and he had no flood arch, whereas these spans are 60 and 45 feet and there is a neat clapper bridge on the left bank to take flood water. The modern bridge, built in 1933 by Mears and Carus-Wilson, spans the river at a different angle with an arch of 90 feet and a small flood arch on the west bank. It is built of reinforced concrete faced with rubble. There are refuges in the abutments.

From Invermoriston it is a pleasant drive up the A887 to the west coast. The bridges on this road are described in Chapter 10.

The next town is Fort Augustus where the Caledonian Canal lifts up towards its summit at Loch Oich and runs beside the river. Both before, and for many years after the canal was built, the lochside road crossed the river on a bridge now closed as unsafe. Old OICH BRIDGE (*NH 381 094*) was first a stone bridge but only one semicircular arch remains and a timber trestle bridge of four spans has replaced the fallen arches. It has rather an old world air with pines and ivy growing in the stone crevices and with the patterns of the wood trusses against water and sky. When I was there on a spring evening a mother and son were fishing from the rickety structure. The main road now crosses the river by the new Oich Bridge built in 1934 by Mears and Carus-Wilson. It has a large, low central span and two smaller ones with stepped cutwaters.

10   Old Fort Augustus Bridge, a combination of stone and timber spans

At one time a railway ran from Invergarry to Fort Augustus, and connected with a steamer on the loch, but it was never a success. It cost £339,000 to build and had a revenue of about £2,000 a year. Opened in 1903, the spur to the pier closed in 1907 and the whole line in 1911. However its closure was seized upon as an issue in the 1912 General Election with headlines such as 'Fort Augustus Isolated' in the *Dundee Advertiser*. The Board said it had failed because local people did not use it. A dealer offered them £28,445 for its scrap value; the North British Railway £22,500. The County Council made up the difference and the line re-opened in 1913, although it had made money while closed from the letting of the embankments for hay crops. It finally closed in 1933, except for a weekly coal train which ran until 1946. At Fort Augustus it came across the River Oich on a viaduct of which only the central pier and abutments remain to become a perch for seagulls.

11   A bridge no more: viaduct pier over the River Oich at Fort Augustus

You can here take a short detour on the A862 towards Foyers to see Whitebridge and the Killin valley.

WHITEBRIDGE over River Fechlin	1732	General Wade	*NH 489 154* beside the A862 at Whitebridge

Many, various and far flung are the Wade bridges pointed out to me in places where no military road would ever have had reason to go, but this is indisputably a Wade bridge on the road he drove from Fort Augustus to Inverness, thus linking two garrisons. For an army bridge it is surprisingly decorative. The first impression is of the height and grace of the arch which is accentuated by the wide approach splay of the parapets. The string course, voussoirs and cartouche are of sandstone, and the string course in particular adds to the harmony of the design by articulating the upper curve of the bridge, while the cartouche makes a point of the high centre. Even the voussoirs are decorative with alternating single and double stones. Like almost all Wade's bridges it is a semi-circle. The span is 40 feet and the road only 8 feet 6 inches. The parapet, 41 inches in the centre, tails down to 20 inches. The abutments are set on the rocks of the gorge with solid tooled stone footings.

This lovely bridge cost £150. It was damaged in the notorious floods of 1829 but well repaired. Although Wade designed it—whatever that means in terms of a busy CO with many other duties on top of road

12   Whitebridge on the old military road, with new bridge behind

building supervision, and it probably means only that he approved and perhaps modified a mason's plans—it was Major Edward Caulfield, his assistant, who superintended the building.

Several cars and a coach stopped to admire the bridge while I was there, which was good to see. Cars whirl past so many beautiful and historic bridges which, if attention were drawn to them, could be enjoyed; and this might result in their being saved for future travellers to see. It is sad to note the many old bridges, retained when a by-pass was put in, are to some extent neglected and will inevitably decay into ruins. Even here a large birch sapling has been allowed to grow out of the cartouche and the voussoir ring is not complete. I am sure this is not lack of interest on the part of the official civil engineers, who in my experience are enthusiastic for bridges. It is lack of money.

The short road up the Fechlin valley to Loch Killin is worth taking for the views it gives of the river's cleft seen in constantly changing perspectives, but you can also see KILLIN CLAPPER BRIDGE (*NH 514 116*) over an un-named burn that falls dramatically away down to the Fechlin as it snakes below the bare sweep of hillside. To accommodate the corner and the slope, the downstream face of the bridge is widely curved, and the upstream face less so. There are three spans, only the central one taking water normally, and, where the stream spills out, an enormous rock has been placed with two flanking rocks, no doubt to direct the flow in times of spate. There is a parapet on only one side and almost certainly the bridge at first had none. The flags were possibly covered with more stones and brushwood to make a roadway but, in stormy weather, must often have been stripped bare. The measurements of various heights and widths involved are all 2 feet or thereabouts. It is this use of roughly square shapes which gives clapper bridges their solid, sturdy lines. Of course the spanning flags have to be considerably larger than the gap they cover. The main one here is 34 inches. Looking inside, you can see that the roof has begun to cave in, which enables you to see the construction, but is sad. There are few enough of these old bridges remaining. To be fair, this one is rather crude. There is almost no attempt at coursing the stones or even of sorting them into sizes, but it may have been inexpertly repaired in the past. The burn is taken under the new road in a corrugated iron tunnel and this has created a fern and reed fringed pool below the clapper where once the burn must have leapt to the river.

Returning to Fort Augustus, you should continue down the Great Glen to Aberchalder where there are three bridges close together over the river and the canal.

The half-ruined OICH SUSPENSION BRIDGE (*NH 338 036*) was built by James Dredge in 1849, and is a type that he pioneered—half suspension and half-cantilever—at the time when engineers were debating and experimenting with various forms of the chain bridge. Dredge's answer was a double suspended cantilever stabilised by the deck held in compression between the abutments with the upper chains giving the tension. These chains were of flat wrought iron rods which reduce in number as the centre is reached. In later bridges Dredge reduced the chains still more because, he said, the weight was not chiefly carried on them but on the towers which are the cantilever piers. The stay rods are diagonal to the tower on both sides, which underlines the method of support.

It makes an attractively romantic contrast to the other bridges beside it, with its solid towers of dressed stone with a pedimented arch between them which make the ironwork appear fragile. It is tufted with windblown plants and the wooden deck sprouts young trees.

The three span concrete bridge taking the road over the river on a rather steep incline was designed by Mears and Carus-Wilson in 1932. It has unusual pointed arches, with a span of 63 feet each, and triangular cutwaters stepped back to the piers. The parapet is shaped in vertical panels which is a sort of modern concrete equivalent of castellation and works well. Though so much younger than Dredge's bridge, it is darkly coated in places with moss. Over the canal is a swing bridge. On the east side you can see the turning gear under the lock keeper's cabin, and also the massive concrete bed it rests upon when opened.

13   Masonry towers of Oich Suspension Bridge, showing cables and rods

SPEAN BRIDGE (*NH 222 817*) was built by Telford in 1819 but was widened in 1932 and so altered that the lines of his bridge are gone. At this village there is an interesting detour up Glen Roy to see a natural arch, if you are prepared to walk.

Glen Roy is famous for its parallel roads—three huge shelves which run round the enormous curve of the valley at a considerable height. They are level beaches left by a lake which once filled the valley and stood at different times at three different depths. It was held in by a stopper of ice, formed by a glacier. They do look uncommonly like roads and have given rise to wild fantasies. Telford and Southey came to see them. Telford said they could not have been made by water because 'the roads had been made since the ravines'. Southey believed 'the roads were intended for a display of barbarous magnificence in hunting'—which shows how even great men can err.

To find the natural bridge, which is a true bridge of arched stones crossing a burn, with a pathway of earth and grass, made not by men but by the chance operation of nature, drive to the end of the road and park where indicated at Brae Roy. Then follow the road on foot over the first two bridges, one a stone arch over the River Turret, the other a plank bridge over the Roy. There is a track both sides but you want the left bank. At a point where the track bends away from the river towards a cottage with three skylights, leave it and keep along the riverside, where walking is easy on the raised bank. Here I saw oyster catchers nesting in April. You come to the Falls of Roy and here must ford the Canal burn which comes in at this point (wellies or bare feet). The Roy was once crossed by many bridges but in 1947, after a very cold winter, huge blocks of ice came crashing down with the thaw and destroyed the piers of 17 bridges in an hour. Few have been replaced. The abutments of one are near the Falls.

The next stream is the Burn of Agie which enters the Roy through a narrow neck. You must follow it up where it loops across the flat land in a reversed S-bend. Ahead you will see a cluster of trees in a rock defile. There is the bridge. Follow the sheep tracks up the hillside and you will know when you are nearing the place when you see the iron stanchions of an old fence. It is in a precipitously-sided gorge, not easy to spot. Once located, it is reached on a narrow path, but take care if you have children with you as the sides are steep and can be treacherous in wet weather.

Your first impression may well be of the fantastically cut and moulded rocks, grey and black, standing on edge, with ribs and hollows like modern sculptures, and the way the clear greenish-white water pours under the bridge and flings itself into space. If you can make your way upstream on either side to a good vantage point, you will see the

14   Natural arch in Glen Roy

15   Estate Bridge in Glen Roy

rocks on the river sides have split naturally into huge cubes and rectangles. It is these that have formed the bridge. One block has fallen, and lies probably on the lip of the fall, and the one above has jammed between its two neighbours thus forming a natural arch of three massive voussoirs. Because of a bend in the stream the arch looks more like a cave and it is hard to find a position in which you can see right through.

The walk, there and back, is about 5 miles. It is full of interest and beauty, and you will have in your ears all day the ceaseless noise of the water.

Just below the car park at Brae Roy is an ESTATE BRIDGE (*NN 331 909*) over the Roy. This wood and metal bridge set down below the field level and creaking in the wind, frames a view of rough water and wild hillsides. It is typical of the bridges built on Highland estates for the use of the workforce, particularly the shepherds, and of course for the sportsmen.

It has concrete piers set into the rock of the river bank—the left hand is almost sheer—and supporting iron girders on which are set a 4 feet plank deck. The sides are of wire netting topped with wooden rails and there are wooden stabilisers at approximately 6 feet intervals, and cable stays on the downstream side fixed into the bank. It is 16 yards long. For the control of stock the entry on the field side is funnelled between fences and the exit on the road side has a gate. There is a set of stone pens in the field.

The rocks in the river bed here are especially fine, a white-grey Leven schist almost like marble, while the smooth slabs on the left bank are a beautiful greyish green. More often than not this is a bleak, wet, windy place where shepherds have tended their sheep perhaps so accustomed to the rounded hills, snow spattered, rising above them with their curious roads that they do not merit particular attention.

Between Spean Bridge and Fort William, on the last section of the Caledonian Canal, are several bridges both unusual and attractive.

HIGH BRIDGE over the River Spean	1736 General Wade	*NN 201 820* off the A82 a mile south of Spean Bridge

About a mile down the side road, is a left bend in front of a modern timber house. Here you can park and there is a path to the near side of the house that leads to the river. The ground is often wet and the path arrives at the broken bridge without warning so, if you have children with you do not let them run ahead.

High Bridge was one of the major achievements of the then Lt Gen

16   High Bridge, photographed in the 1930s and already a ruin.
Courtesy RCAHMS

Wade's military road building scheme undertaken for the government between 1724 and 1740 and carried on by his successors. Because the river cuts down in an extremely steep valley, Wade decided to build a high bridge so that access was not too difficult, though it is steep enough as you will have seen. He built a three span bridge of widths 40, 50, 40 feet. They were to be 80 feet above the water, which is 30 feet deep, so that they have an overall height of 110 feet. The bridge was his standard width of 16 feet, and 280 feet long. It cost £1087. 6s. 8d. which was a considerable amount.

It is difficult now to visualise High Bridge as it was in 1736, especially as a section of iron truss is perched across part of the gulf. Trees grow out of the masonry and thickly up the gorge sides. Yet, even now, it is an impressive piece of building and must have been doubly so when first completed in a land where roads were a rarity and at a spot where previously only the bravest swimmer would have dared across. No old ford here. The black water pours below carrying swirls of foam flecks.

Historically the place is of interest because the first skirmish of the 1745 rebellion took place here when two companies of Royal Scots going south were met by twelve men of Keppoch's under Donal MacDonald. Two redcoats were seized and the rest fled. The line of Wade's road can be followed from here. The bridge at Nine Mile Bridge in Glen Gloy is one of his but so defaced by later repairs it is not an inspiring place to visit (*NN 224 864*).

MUCOMIR BRIDGE   1813   Telford   *NN 184 839*
over River Lochy   take the B8044 off A82
                                                    at the Commando
                                                    Memorial

Before 1813 the River Lochy did not flow into the Spean here but
farther west at Gairlochy. Telford altered its path because he wanted to
run the canal through at Gairlochy. A new rock channel was cut which
made the river deeper and not fordable. However the drovers used this
route to the south so Telford had to provide a bridge. Somewhat
ironically, the displaced river has been altered again by later engineers
who have utilised its flow for a power station. One of the main arches
now spans the turbine race, and the river bed below has been deepened,
while the other arch spans the sluice spillage and here a raised bed has
been constructed. The profile of the bridge today is shown in Figure 11.

Telford's handsome bridge had spans of 52 feet and 29 feet. It was of
coursed rubble with a sandstone string course and voussoirs, and with
triangular cutwaters below semi-circular spandrel buttresses each
decorated with cross-shaped and vertical slits. The stone came from the
banks of the Spean. Later, and no-one knows exactly when, a third arch
was added of different design and materials. No water runs under this
now and it is largely hidden by scrubby trees. The present bridge is 88
yards long and has a slope of 1 in 23.

17   Mucomir Bridge over the re-positioned River Lochy

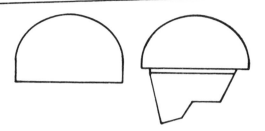

Fig 11   Mucomir Bridge, profile after alteration for hydro scheme

In 1960, when the power station project was beginning and heavy lorries had to be thought of, the engineering firm of Crouch and Hogg stripped down the arches to the barrels and filled them with reinforced concrete. A thrust slab was inserted below the spring of one arch. To my mind this detracts from the appearance of the bridge. There has also been damage to the string course and some ugly cementing. However the lines of Telford's bridge can still be appreciated. I visited it on a February day of fierce bright wintry sun which came in sudden flashes between heavy cloud and downpours of cold rain. The mountains were white and Ben Nevis knife-edge clear yet there was a feel of spring. The hazel catkins were yellow and the alder twigs fox red.

If you continue along the road to Gairlochy you will see, on the left, a sump-like hollow which is the old course of the Lochy. In the calm of today it is hard to imagine the confusion and energy of 180 years ago when this whole area must have been a works site. The heavy draught work was done by horses who pulled three ton loads of stone and worked a ten hour day. The stable for them here cost £178. 4s. 4¾d. to build. It was hard work for the men too, often in bitter weather. One year the canal accounts included the sum of £127. 7s. 0½d. for whisky given to men toiling in the water.

MOY SWING BRIDGE   1812   Telford,   *NN 162 826*
over Caledonian Canal   Thomas Rhodes   along the towpath
                                                          from Gairlochy

This charming bridge of painted cast iron is the only one remaining of the original swing bridges over the canal. They were probably all like this at first.

The iron was cast at Plas Kynaston by 'Merlin' Hazledine who made all the iron work for the canal, and it was constructed on site by Thomas Rhodes. The bridge swings in two halves. The teeth and pinion mechanism is below on the side of the canal. The handles are in two

18   Moy Swing Bridge, the only remaining original bridge over the canal

short metal pillars. It is 30 feet long. The railings splay out like banisters and there is a half-moon curve fixed to one end to fill the gap over the water when the bridge is open. The gate on the far side has an unusual hinged bolt.

When first built it had no railings. They were added in 1850 after the accidental death of a contractor called Bean who was constructing a second lock at Gairlochy and was thrown from his horse as he crossed this bridge. Now the diagonals seem an integral part of the design and of the pastoral scene, the white cottage, shining water, pines and great hills.

A section of the railing was damaged by a boat in 1988 and late in 1989 had not been repaired, which neglect of a historic and pretty bridge is a pity to say the least.

Because there are two halves to open, the keeper has a boat. In summer as many as sixty boats pass in a day, so he leaves the bridge open and closes it when a farm vehicle wants to cross. The canal traffic is mostly pleasure craft. Gone are the days, remembered by an old lock keeper who lives nearby, when upwards of fifty herring drifters were waiting on Monday mornings to enter the canal at Corpach. The keeper's cottage, which completes the charm of the place, has windows facing both ways, as in a toll house, and even a small window in the gable so that he could watch the water from his bed.

At first Telford and his assistant in this area, called Easton, considered building an aqueduct here because they feared the flood force of the Moy burn. Eventually they contained the burn another way. If you walk 100 yards towards Gairlochy to where the burn enters the canal and flows out again, you wil see the small five-arched bridge built to take the farm track over. Up the burn are waterfalls constructed to slow its rush, along with pools to trap boulders and gravel which it might bring down and which must not be washed into the canal to silt it up. On the far side a similar outlet span has partly collapsed, and the reservoir beyond been allowed to drain.

GLEN LOY AQUEDUCT over the River Loy	1806	Telford, Alexander Easton, Simpson and Wilson	*NN 149 818* where the B8004 crosses the Loy

It is easy to miss this superb aqueduct as you pass, so snugly does it fit into the canal bank and the surrounding farmland, but as you walk towards it on the track which disappears into one of its side tunnels, you appreciate the ponderous beauty of its massive curving front. The wall of masonry tends to dwarf the three arches, one for the river and two for farm traffic, to such an extent that you are surprised to find how large they are.

This is a very different thing from the famous aqueduct built by Telford at Pont Cysyllte which was called a stream in the sky. Here the problem was to carry a wide, deep canal, built for sea-going ships, over a small but fierce mountain river. The tunnels are 84 yards long, which gives some idea of the size and weight of the water-filled canal above, and their widths are 10, 25, 10 feet. They are 8 feet high and have triangular cutwaters. Telford's deputy, Easton, was in charge; the contractors were Simpson and Wilson, while the foreman was named Cargill. The work was completed in one summer with one hundred masons working on it. The timber battens and trestles needed were made on site for there wa a saw mill at Strone. The stone was brought from Banavie and the granite facing from near Ballachulish.

If you walk through—beware of puddles—you will notice the huge rectangular granite stones, salmon-orange in colour, that floor the tunnel, and you can hear the river running in the parallel tunnel. On the far side the stonework has been repaired with concrete. Above, the canal takes a long, flattened curve, serenely ignoring its footworks.

Telford's engineers also built the road alongside the canal which included fifty-six single arch bridges in a 10 mile stretch, since it has to cross many burns which rush down the hillsides in rainy periods. No designs were drawn for these bridges; the masons were merely told the dimensions to be put on a standard pattern. They had to be 16 feet wide and have a gravel surface to a depth of 14 inches. The gravel stones were to be at least as big as a hen's egg and the inspectors carried gauges to check. The bridge over the Loy where you parked is one of these, since strengthened.

At TORCASTLE (*NN 132 791*) there is another aqueduct worth visiting. You can park near the telephone kiosk and walk down the track, bearing right. This one has three equal-sized arches and the pedestrian tunnels slope down and up again. The spans are approximately 9 feet and the height is 13 feet—large enough to take a car as I discovered to my surprise when placidly photographing. From the towpath above there is a wonderful view of Ben Nevis. There are two other aqueducts on this stretch of the canal, one near Muirshearlich (*NN 144 805*) and a second at Mt Alexander (*NN 122 777*), neither of which I have seen.

This road will take you on to Neptune's Steps and then to Fort William, where, within a mile of each other, are two railway bridges both built for the West Highland Extension Railway to Mallaig.

The LOCHY VIADUCT (*NN 119 755*) is the earlier, built in 1895. It is a four span steel truss bridge, each span 74 feet long and 20 feet high, carried on stone piers and abutments given castellated decorations. These rather baronial towers are only semi-circular above the deck in order to accommodate the rails. It is worth visiting for the site alone. (Take the turning to Caol. You can park down a side turning nearby.) It is a scene of contrasts—the sturdy, grey-painted viaduct with its

19   Torcastle Aqueduct under the canal, with traffic

Victorian hint at medieval castles, the narrow footbridge with its locals trailing dogs and shopping bags, and behind, first the romantic ruined towers of Lochy Castle, then the bulk, power and loneliness of the mountain.

BANAVIE SWING BRIDGE (*NN 112 768*) crosses the canal and was built in 1901 by Simpson and Wilson. It is a steel hog-back bow truss, painted cream. The bow is not symmetrical, being more heavily braced over the turning mechanism. An inscription in the metal says it was cast at the Parkneuk Works at Motherwell by Alexander Findlay and Co. The original wooden signal box has given way to a very technical looking replacement which is in stark contrast to the rotting timbers of the bridge's bankside supports and to an air of partial neglect.

A short distance further on at Corpach the canal enters Loch Eil and the very last bridge over it is the one on top of the final lock gates. It was here that Southey wrote his eulogy to the canal and his friend Telford. He wrote of the canal which he considered Britain's greatest work of art:

> The pyramids would appear insignificant in such a situation, for in them we should perceive only a vain attempt to vie with greater things. But here we see the powers of nature brought to act upon a great scale, in subservience to the purpose of man: one river created, another . . . shouldered out of place, and art and order assuming a character of sublimity.

III

Fort William to Connel

Map 3

At Fort William the Mallaig road, A830, leads westwards following for the most part the line of Telford's road built in the first years of the nineteenth century. Four miles from Corpach there is a signposted turn to FASSFERN BRIDGE (*NN 021 789*), a small, low-set bridge with noticeably curved spandrels, as the early bridges have. The span is 24 feet and the abutments are founded on the rocks. This bridge should be seen in conjunction with DRUM NA SAILLE BRIDGE (*NM 960 794*). At the junction of the A861 with the A830 at the western end of Loch Eil a short turn to the right leads directly to this small bridge. Telford instructed the masons to build in four distinct, but related, styles. The specification for the first type said that the parapets should curve horizontally—that is they rise from almost ground level to a highest point at the centre of the bridge—and that they should also curve vertically—that is they are slightly battered. Moreover, the spandrel walls were to have a concave curve of 4 inches. Add these curves to the arch of the vault and you have an extremely complex shape which is a pleasure to look at but may well have been very difficult to build, particularly where the two banks of the stream were not at the same height or the ground was not level. This pattern seems to have been almost entirely restricted to smaller bridges and fairly soon gave way to the second type which had modified curves—the parapet top became horizontal and the road more nearly level. Drum na Saille is an interesting example of the first type. Fassfern is similar though larger.

GLENFINNAN VIADUCT	1901	Simpson and Wilson	*NM 910 814*
over River Finnan			park at the Memorial and walk ½ mile up estate road

This magnificent curving bridge was built for the West Highland

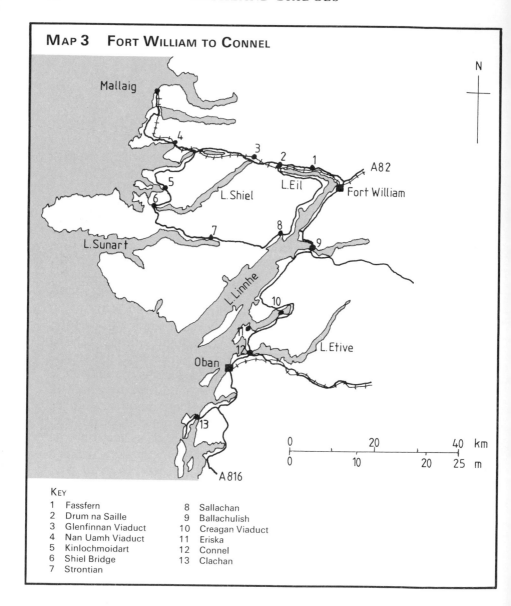

MAP 3   FORT WILLIAM TO CONNEL

N

Mallaig

A82

L.Eil
L.Shiel   Fort William

L.Sunart

L.Linnhe

L.Etive

Oban

0   20   40 km
0   10   20   25 m

A816

KEY

1   Fassfern   8   Sallachan
2   Drum na Saille   9   Ballachulish
3   Glenfinnan Viaduct   10   Creagan Viaduct
4   Nan Uamh Viaduct   11   Eriska
5   Kinlochmoidart   12   Connel
6   Shiel Bridge   13   Clachan
7   Strontian

20   Glen Finnan Viaduct

Extension Railway and is still used by trains to Mallaig. It is constructed of mass concrete and is the longest concrete railway bridge in Scotland. It has twenty-one semicircular arches at a height of 100 feet and an overall length of 416 yards, while the curve is 284 yards. Only five of the piers straddle the river and its flood plain, and just one actually stands in the water and is thus provided with a cutwater. This centre section of a long bridge is articulated by the two outer piers of the five being at least twice as wide as the rest, which are, in fact, remarkably slender. (They gradually get shorter of course, as they climb the valley sides.) In the parapet there are small square refuges over every second pier and some of these have the parapet cut away on each side; some not. I could find no symmetry in this. Apart from the refuges and a small tiled sloping ledge at the spring of each arch, the viaduct has no decoration and needs none. The simplicity of its shallow curve perfectly fits the massive outlines of the mountains. Behind you can see the small road winding uphill into the grey and green hills. At its feet stands a solitary pine and a wooden bridge to serve the small community here. Rowans and birches grow by the river and in September scabious and the orange seed heads of bog asphodel were thick in the yellowing grasses.

It is a remarkable place. Here is this immense bridge on its elegant concrete legs built nearly one hundred years ago when the railway network appeared capable of almost infinite extension; and below is the memorial to an entirely different age when some Highlanders still planned for a separate and independent state in which the clan was the unit of power rather than the commercial politics which builds viaducts to remote ports.

A further 14 or so miles west the road passes under another West Highland Railway viaduct the NAN UAMH (*NM 729 841*). It stands over the Leann Mama where it enters the sea loch of Nan Uamh across a pebbly beach. It is a smaller bridge with eight segmental arches of 50 feet span, but has a pleasing solidity. Its arches frame views of the sea and islands and when I was there cattle were grazing on the beachside grass. It also was built in 1901 and designed by Simpson and Wilson. The piers are thick and the central one very wide and slightly stepped out from the façade. This detail is matched on the abutments.

Those interested in railway architecture may wish to visit BORRODALE BURN VIADUCT (*NM 698 855*). It has three spans, the central one being 127 feet and, when built, it was the largest mass concrete span in the world. I have not visited it.

To continue towards Ardgour, where you can pick up the ferry and continue the drive towards Connel, you should return to Lochailort and take the A861. The name of the Rough Bounds was given to the Moidart peninsula in the eighteenth century and even now it is a remote corner of Scotland. Once off the few roads, you are in a wild area of steep hills, innumerable burns and a thousand lochs. The road for the most part skirts the coast, and was first built by the Parliamentary Commissioners in the early part of the nineteenth century. It must have been a formidable undertaking, but they believed in making roads where people needed them, and not where a military strategist might place them. To the isolated coastal communities, who had until then relied on boats, the new road must have resulted, perhaps only slowly, in a complete reorientation of their lives.

At Ardmolich stands the lagest of the bridges on this early road— KINLOCHMOIDART (*NM 712 721*). It spans the River Moidart at the head of the loch. It is built to Telford's first bridge design with the double curvature of the parapet and spandrels but because of its size, the span is 51 feet, the curved effect is somewhat modified. The abutments are wider than the bridge itself and the spandrels curve round to join them. The arch is a flattened segment with many of the voussoirs almost vertical. The rough stones coping the parapet overhang the spandrel wall.

It is a lonely and beautiful place. There are just two houses and a stone shed with a corrugated iron roof. The river snakes through the lush meadow grasses and the hills at hand are covered with trees.

SHIEL BRIDGE (*NM 677 689*) is a very different structure. Built in 1930 to by-pass a fine old Parliamentary bridge, it spans the River Shiel

21   Kinlochmoidart Bridge. Notice the flattened arch and vertical voussoirs

where it flows through a very boggy, flat valley as it leaves Loch Shiel.
The bridge has three segmental spans and the central one, at 14 yards,
is wider and taller than the others. It is built entirely of dressed stone in
the rustic ashlar style and the circular cutwaters extend into rounded
piers which finish with refuges. Over the abutments the refuges are
square and the whole parapet is castellated. It has a slight rise to the
centre and on the north side there is a separate flood arch linked to the
bridge with a railing. If you look over the parapet at the footing of the
piers you will see, if the river is not flooding, the stone starlings built to
protect the piers from damage.

One might anticipate that such a mock baronial bridge would be out
of place in this remote spot. However the bridge, not truly handsome
and lacking charm, does have sufficient good looks to fit into the land-

scape. This is due chiefly to the way the arch curves down to the low set cutwaters and the general neatness of the design. I have not been able to discover who the engineer was.

At Strontian the road bridge is a single span modern one with a stone facing to the concrete that is not coursed. This gives a false naive look which seems a particularly odd style to adopt in an area with so many fine coursed rubble bridges. The old bridge at Strontian is a case in point.

Take the side turning to the church. This is the start of a beautiful road running eventually through to Polloch on Loch Shiel and passing on the way the remains of the large scale lead mining that used to go on here. STRONTIAN BRIDGE (*NM 817 625*)—date uncertain, but perhaps in the late 1700s—is a goodlooking hump-backed bridge in a peaceful and fertile stretch of the River Strontian. It is built unusually of dressed stone; not dressed with great finesse or precision but not the usual rough blocks either. It has two elliptical arches of 20 feet span and a flood arch. The triangular cutwaters continue up into the spandrel walls as narrow buttresses. The neat parapet edge overhangs the spandrel and the abutments are buttressed and have a wide splay. The whole bridge is slightly curved to the south facing Loch Sunart. It is in somewhat poor repair.

SALLACHAN BRIDGE (*NM 978 628*) spans the river draining out of Loch nan Gabhar and into Loch Linnhe. It also is by-passed. It is a hump bridge with two segmental arches of 27 feet, which means that it is quite long and this results in a curious profile with a flattened top, the road rising to the first crown and then running level to the next before sloping down. This can best be appreciated at the water side. The river here is wide and slow, a fine stretch for fishing. The bridge is of coursed rubble with narrow voussoirs and rounded cutwaters possibly later additions. When I was there in September, the sun was just going down behind the purple mountains of the Ardgour peninsula and casting a delicate, softening rose light on the water and hilly banks of Loch Linnhe. From the coast I had a spectacular sunset view of the Ballachulish inlet, with the mountains of Glencoe all immersed in a madder and violet sky. The focal point of this sleeve of mountains, water and sky was the bridge at Ballachulish, which shone a pearly cream and seemed the finishing touch the valley needed, underlining its shape. I wished those who castigate this rather severe bridge could have seen it then.

BALLACHULISH BRIDGE 1975 W A Fairhurst and Ptnrs *NN 052 598*
over the mouth of Loch   Cleveland Bridge   junction of
Leven   Engineering Co   A82 and
   A828

Ballachulish has always been an important junction for travellers and there are few motorists who do not know it. Before this bridge was built it was a name to make them groan at the thought of either a long delay waiting for the ferry, or an even longer detour up to the head of Loch Leven. Now it can be sped through, which is a pity because there is much of interest, and there can hardly be a bridge with a more lovely view into the mountains, or out towards the sea; but, for such a place of romance and history, the bridge seems to some people out of place in its matter-of-fact appearance.

It is a long, plain steel truss bridge which is painted a dark olive that blends well with the hillsides and, as someone said to me, 'In certain lights you can't see it.' I am quite happy to see it. It is a business like construction set here to do a job and not trying to make a romantic statement.

The cantilevers are supported on abutments set well back from the loch. The road rises slightly from there to the piers at the water's edge. The piers are narrow rectangular columns and the studs on which the truss rests can clearly be seen over the water. The bridge has an unusual profile with a slight hump over the piers and a raked arch over each

22   Ballachulish Bridge, with maintenance work in progress

entry. The beams of the trusses slope inwards towards the piers so that there is a quiet but definite geometrical pattern. In fact, understatment is the chief characteristic of this bridge which does not try to compete with the magnificence of the scenery but unobtrusively gives you passage through it.

It can be viewed well either from the old ferry landing or from the Oban road which passes under it. Here steps lead up to the bridge and you can walk over it and admire the views, the boats riding at anchor or sailing out to sea, perhaps a heron fishing at the edge. From here the sturdy construction of the metal work is evident. I find the lights some-what out of keeping. To provide safe, unobtrusive lighting for their bridges seems to be a problem for designers.

Just above the stairs is a memorial to James Stewart who was hanged in this spot in 1752 for the murder of Colin Campbell of Glenure. His guilt is still disputed. The event was used by Stevenson in *Kidnapped*, a marvellous story set largely in this part of Scotland. He draws an unforgettable picture of Allan Breck who was Stewart's ward and another suspect.

CREAGAN VIADUCT over Loch Creran	1903	Sir John Wolfe-Barry and others; Arrol's Bridge and Roof Co of Glasgow	*NM 978 443* at Creagan beside the A828

Like Ballachulish, Creagan tends to be remembered by motorists for irritation caused by the detour round the head of Loch Creran. The viaduct spans the narrowest point of the Loch. It is rather a fine truss viaduct built for the Ballachulish extension of the Callander to Oban railway in 1903 and closed by British Rail in 1966. The main engineer involved was sir John Wolfe-Barry, but he worked in association with Messrs Forman, McCall, H H Brunel and E Crattwell. The bridge contractor was Arrol's Roof and Bridge Company of Germiston and the railway conractor was John Best of Edinburgh.

This team has clearly tried to fit the modern steel bridge into the Highland scenery by providing castellated granite abutments and pier. Whether this approach is better than the out and out functionalism of Ballachulish and Connel is a matter for debate but I am inclined to think not.

The bridge has four spans, two outer ones of pointed masonry arches and two inner ones of Pratt steel trusses. There is a cantilevered walkway on the east side. The road passes through the side arch on the south and this arch has a brick vault. The pier has a low arch at water level—an unusual feature. There is a denticulated string course and the

23   Creagan Viaduct. Note the low arches on the pier

abutments have tall outer portals of half towers. Altogether the design is rather heavy. The tidal water flows strongly and smoothly past and there is a pervasive smell of seaweed. Inland, the tops of the mountains are both sunlit and mist-veiled tempting the exploration of Glen Creran and Glen Ure on foot.

The Scottish Development Department, to whom this bridge now belongs, have researched to discover documents that would reveal what foundations the bridge has because they wish to convert it for road use. The north abutment and pier are on rock, the central pier is on two steel caissons; but the foundations of the southern abutment and pier on the alluvial plain are not clear. Almost all the technical drawings have gone, and the bridge is barely mentioned in articles on the railway. This is something I have come up against many times in writing this book—records of comparatively modern bridges are lost. It is often the case that eighteenth-century structures are better documented than those of the twentieth century. This is a pity, for today's modern engineering will be tomorrow's industrial archaeology. I like to know the name of the man who built a bridge I use with gratitude; and would urge County Road departments to preserve such human details along with technical matters in their computerised records.

After Creagan, at the mouth of Loch Creran a peninsula projects which has an island called Eriska off its northern tip. A narrow channel called An Dorlinn flows between the two and is crossed by ERISKA

BRIDGE (*NM 898 424*). It is a two span cast iron lattice girder bridge and its extended abutments have square sea flood holes. It is painted a bright dark green and has ornamental gates at the south end. The scenery in May was perfect with hawthorn in bloom above bluebells and seapinks, and with the spring foliage against blue waters, blue sky and the bluer mountains.

CONNEL VIADUCT over the mouth of Loch Etive	1903	Sir John Wolfe-Barry and others; Arrol's Bridge and Roof Co	*NM 911 343* at Connel on A828

The extension railway from Connel Ferry Station to Ballachulish was 28 miles and cost £400,000, more than had been estimated because of the difficult terrain. The bridging of the Falls of Lora at the mouth of Loch Etive was particularly difficult. There is a tidal drop of 4 feet across a stretch of water 350 feet wide and the current is rapid, perhaps 10–11 mph at a spring tide, and the water had to remain navigable. So a cantilever construction was decided upon. The two main granite piers are 524 feet apart and the clear span is 500 feet—the second longest clear span in Europe at the time. (The Forth bridge was longer.) At highwater the height to the underside is 50 feet and to the top is 125 feet. It was built by the same team which built Creagan, led by Sir John Wolfe-Barry.

It is a bold and dramatic construction best seen at the waterside. The main features are the huge triangular cantilever supports. There are two

24   Powerful diagonals on the Connel Viaduct

legs, on a masonry pier, jutting forward and upward. These legs measure 6 feet 7 inches by 4 feet 6 inches. They are tied back to the abutments by the upper members of the trusses and by back struts. The large triangles thus formed are bisected by the deck so that two smaller triangles show. Then the struts are strengthened with diagonal girders forming further triangles, and the two legs have diagonal struts. The granite piers are low and oval, 24 feet wide and 110 feet long. Founded on rock they were built within cofferdams in the shallow water. Each was built for a maximum load of 1,500 tons. The abutments, into which the truss girders are tied by anchors 50 feet below the deck, extend on each bank into three semicircular masonry arches. The bridge used 2,600 tons of steel and the workforce averaged two hundred men. Housing them in what was then a remote area and during a particularly severe winter was a problem.

When finished, the bridge was inspected by a Colonel Yorke on behalf of the Board of Trade. He arranged for no less than nine Caledonian engines and four 30-ton wagons, all loaded, to cross the bridge stopping frequently. The load was about 1,000 tons. This train then crossed several times at increased speed. The entire line was officially opened on 21 August 1903 by the passing of a special train from Buchanan Street in Glasgow carrying railway and engineering men. A party was held in a hotel at the Ballachulish end. In the early years five trains a day crossed the bridge. Later it was used for road and rail. Now it is a single track road.

Driving along the bridge (if you are a passenger and need not concentrate on the road ahead) you are aware of moving through a corridor of changing cross-over patterns and angularly framed views. On the south side the old railway approach embankment can be seen.

An addition to this route down the Great Glen to the Atlantic is an off-shoot drive through Oban to see the only bridge which spans that ocean. Once past Oban, take the A816 as far as Kilninver where the B844 turns off.

CLACHAN BRIDGE over	1791	John Stevens	NM 785 197
Clachan Sound		(and Robert Mylne?)	on the B844
			at Seil Island

The minor road towards Seil is pretty—in June it has yellow flags blooming in the rushy inlets and bushes of elder flower—but take care as it is a dangerous mixture of dual track and passing places.

The bridge is a steep hump over an unusually high arch which was necessary to allow small vessels to pass beneath. It is 28 feet above the water.

25   Clachan Bridge over the Atlantic

The need for a crossing to the island was discussed in 1787. One proposal was for the narrow channel to be filled in, but this idea was dropped, possibly due to fear of the sea ripping it out. John Campbell of Lochend, who was Chamberlain of the Breadalbane Estate, got plans from John Stevenson of Oban. The cost of this first design was estimated at £450 and it had two arches. It is probable, but not certain, that Robert Mylne, who was working at the time at Inverary, amended Stevenson's plans to one arch and added the oculi. Whatever happened, it seems poor Stevenson made a loss.

The bridge is a segmental arch of 72 feet, roughly coursed of boulder rubble with high wide spandrels in which blind oculi are set, 7 feet diameter. The entire length, with approaches, is 104 yards. It has an unusual parapet which only runs over the actual arch and finishes with a square stone over the oculi. Growing in the walls is a pretty violet flower called fairy foxglove (*erinus alpinus*) which adds to the pastoral look of this pretty bridge in its setting of green banks with copsy trees, an old white farmhouse, boats moored and perhaps clouds piling up over the sea beyond; and it is certainly strange, when walking on it, to see what looks like a river running to the sea in both directions.

IV

The Rivers Nairn, Findhorn and Lossie

Map 4

1 The River Nairn

CULLODEN VIADUCT	1898 Murdoch Paterson	*NH 763 450* seen from B9006 near Clava

One of the most stimulating places to see this viaduct from is the site of the Clava burial cairns, the contrast between the mysteries of ancient man and the thrust of industrial man is so sharp; although it is always possible that some future visitor centuries hence will gaze at what remains then of the railway and at the prehistoric mounds with the same incredulity.

There is no point where one can see the whole viaduct, the longest in Scotland, because it curves away at the south end. It is most clearly seen from the road bridge over the railway just before the viaduct starts. It was built by Murdoch Paterson as part of the final stage of the Perth to Inverness railway which Mitchell built in the 1860s. His line ran from Carrbridge to the coast and avoided Slochd summit by approaching Inverness via Nairn. Passengers agitated for a quicker route and at the end of the century Slochd was finally negotiated.

Culloden Viaduct is 1,800 feet long and has twenty-eight spans all of 50 feet except the central one which is 100 feet. They are bold and simple, remarkably free of decoration. The piers taper upwards and the semicircles are set neatly on them. It is built of bullfaced red dressed rubble with tooled ashlar voussoirs and parapets.

Between the B9090 running past Cawdor and the A96 there are several small roads on both sides of the Nairn. On one of these minor roads near Cantray House is **CANTRAY BRIDGE** (*NH 800 480*) at a point where the road makes a V-bend on a hill. It is dated 1774 and replaces an earlier bridge. It has two segmental spans of approximately 27 feet

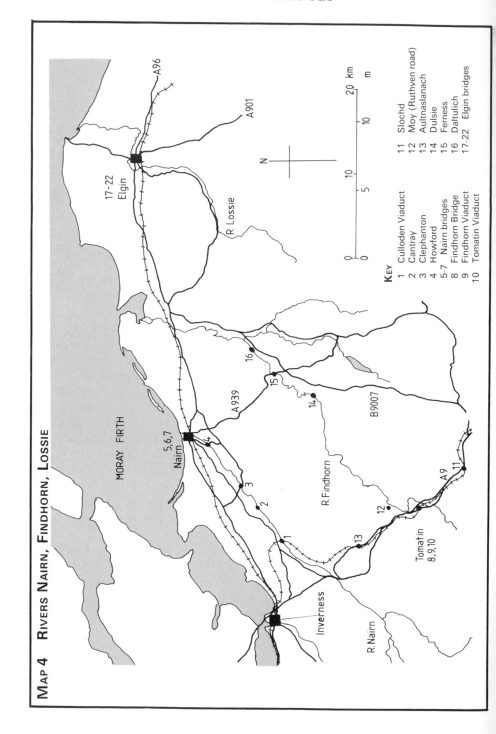

MAP 4 RIVERS NAIRN, FINDHORN, LOSSIE

KEY

1 Culloden Viaduct
2 Cantray
3 Clephanton
4 Howford
5-7 Nairn bridges
8 Findhorn Bridge
9 Findhorn Viaduct
10 Tomatin Viaduct
11 Slochd
12 Moy (Ruthven road)
13 Aultnaslanach
14 Dulsie
15 Ferness
16 Daltulich
17-22 Elgin bridges

26   Culloden Viaduct strides across the River Nairn valley

and a sloping roadway. It is built of coursed sandstone rubble but the coping stones on the unusually high parapet are dressed and of good size and uniformity. On the upstream face is a carved coat of arms, which seems to be of a boy among foliage with a defaced motto. Above is a plaque giving the builder's name as David Davidson, Esquire of Cantray. No doubt he financed the bridge and may have designed it. The mason is unknown. There is a calm loneliness about this place. Above the bridge is a Victorian lodge cottage beside ornamental iron gates, propped open; the house empty. A short distance away is a big farm square, also empty and its clock removed from the gate tower.

Where the B9090 crosses the river is CLEPHANTON BRIDGE (*NH 823 503*). In most respects it is a typical bridge of its period—the mid eighteenth century—with two arches, the larger segmental and the smaller semicircular with a central pier in the river, triangular cutwaters and some modern concrete strengthening. The unusual feature is only to be seen from the riverside. From here the coursing of the stone will be seen to be radial, rather than straight, so that the lines ray out round the arches and, where the two circles meet over the pier, join in a herring-bone pattern.

This bridge is on the line of Caulfeild's military road which comes pretty straight from Dulsie until it zig-zags to avoid Cawdor. It is not however the original bridge. Its exact date and engineer are unknown, though 1764 is hazarded.

27   Diagonal spandrel courses on Clephanton Bridge

When the B9090 crosses back to the left bank of the river it does so on **HOWFORD BRIDGE** (*NH 877 538*) which is built on a slight skew at a point where the water runs over pebble beds and is overhung by alders. It was built in 1905 by P McFarlane Cram—a well-made bridge but somewhat dull. It is a two span box girder truss with diamond lattice rails and curved stabilisers all painted pale green. The rather ornate rustic ashlar abutments and pier have small rounded cutwaters. The Brandon Bridge-building Company of Motherwell made the metal work and it was financed by Nairn County Council, according to a plaque on the bridge.

NAIRN TOWN BRIDGE	1803	unknown engineer	*NH 886 566*
	1829,		on the A96
	1868,		in town
	1936		

The many dates given for this handsome bridge are explained by a tablet on the upstream face which says that the original bridge was damaged by the 1829 flood and rebuilt, supposedly to the same design, by the Trustees for the turnpike road; then, again in 1868 it was 'partially re-erected' this time by the Trustees of the Town and County Roads. The final date of 1936 signifies the biggest change in the bridge, for it was

28   Masonry details on Nairn Town Bridge. Note recessed voussoirs, and
      architrave, cutwater top, oculi, denticulate string course and coping

doubled in width which explains the different looks the upstream and downstream sides have. The parapet was moved but otherwise the first bridge was left and a second matching one put beside it. If you walk underneath on the path provided you can see clearly what was done.

The 1803 bridge had three segmental spans, two over the river and one spanning the riverside path. There is a string course with denticulation; large voussoirs with every alternate one recessed and a moulded extrados; blank oculi circled by nicely shaped rims; and rounded cutwaters with decorated tops. The whole design has a pleasing crisp simplicity and also illustrates neatly many features of bridge design of the period. Over the pier footed in the river a small oval entablature

29   Cycling along the sewage pipe

probably gave the designer's name but now only the word 'architect' is readable in flowery lettering.

Because the road is on a slope the spans increase in size to the west. On the east side the abutments extend into a long wing wall finished off with a string course to harmonise with the bridge. If you walk towards the sea along the footpath you will see to advantage the slope and spring of this handsome bridge.

Also here you will notice an unusual footbridge (*NH 887 567*) carried on a sewage pipe. The abutments are of stone and quite handsomely built suggesting the possibility of an earlier bridge. What is here now is a large pipe riveted in two sections and supported on two cylindrical metal columnar piers. Metal girders bolted to this pipe, and strengthened with angle bars, carry a wooden deck which is 40 inches wide and has stabilisers. It has silver painted railings and matching silver lamps on the abutments.

Nairn makes good use of its river and there are several footbridges over it within the town limits which can make a pleasant stroll. Beside the old churchyard with its huge yews is the impressive railway viaduct.

NAIRN VIADUCT   1857   Joseph Mitchell   *NH 886 562*
                                                             riverside path

This tall, somewhat severe grey bridge curving across the river was built for the Inverness and Aberdeen Junction Railway by Mitchell. It has four segmental spans and a semicircular flood arch on the west side. The piers have semi-hexagonal pillars above them which rise to make refuges at the parapet level. These refuges are corbelled which gives the effect of castle towers, in line with Mitchell's usual style when building railways in the Highlands. He described the work on this bridge as 'heavy and formidable' as, to find the rock foundation they needed, it was necessary to excavate 'a bed of solid gravel between 6 and 10 feet in depth'.

I find the side flood arch is a splendid shallow curve of stone to walk under. The nearby road is taken under the embankment by means of a semicircular arch.

2   The River Findhorn

FINDHORN BRIDGE	1926 Owen Williams and	*NH 804 277*
	Maxwell Aryton	on the old A9
		south of Tomatin

In the 1920s when the Parliamentary north road was substantially rebuilt and re-bridged to become the A9 Sir Owen Williams (1890–1969) collaborated with Maxwell Aryton, an architect, on many of the new bridges. Only a few of these remain in use, and some have been demolished which is a great pity as they are all fine examples of reinforced concrete building of that period.

Seen from up or down the valley, and particularly from the minor road to Raigbeg, this long bridge with its three little flood arches under the approach embankment looks relatively graceful but, close to, I find it rather intimidatingly heavy.

Williams had to cross a valley 200 feet wide and water 30 feet deep. He did so (at a cost of £33,146) by using a concrete truss construction. There are two trusses, each 98 feet by 36 feet, and they meet on a curiously split central pier. Each truss has two beams supporting a concrete deck which is secondarily supported by transverse beams. These are bowed, and correspond in their spacing to the verticals of the arcade under which they protrude lightly as bosses.

Aryton's contribution to the design seems chiefly to have been the arcading of the truss walls. Those who admire this bridge emphasise the skill with which Aryton underlined the basic structure by his detailing. The apertures in the walls are huge and a splayed half-octagon in shape. They rise 8 feet 9 inches from the deck and are 9 feet 5 inches wide at the parapet edge. There are seven in each span, that is to say twenty-eight altogether, and then three more in each of the four bays built over the

30   Findhorn Bridge, from the river bank

abutments. So we reach a total of forty. In addition the central piers each have two downsloping slits in them, something between spy-holes and drains, which mark the separation of the trusses. This arcade is so tall as almost to demand a roof.

In a photograph taken in the 1920s, when the concrete was clean and not pocked by lichen as it now is, the articulation shows more crisply and in greater detail. In a drawing probably made by the architect, this is even more evident. Perhaps Williams and Aryton would be disappointed by its present appearance. On the whole the looks of concrete bridges do not improve with time, whereas a stone bridge becomes an ever more mellow part of its surroundings.

Even although I have, in writing this book, learned to appreciate Owen William's work, I cannot entirely come to terms with this particular bridge although it is admired by some. One writer calls it Williams and Aryton's 'finest joint achievement'. It reminds me of a blockhouse, having something of the heaviness and aggression of military structures. Nevertheless from the water meadows, looking along the underneath of the deck, the construction is certainly impressive.

There is a grandiose inscription which says the bridge replaced Telford's built in 1833. (However, the present bridge did not follow

31   Findhorn Bridge, details of abutment, note size of pedestrian

directly Photographs taken by Major Bruce's team show a two-span metal truss here in 1920.) On the inscription the names of the Convener of Inverness County Council and the Minister of Transport are cut in large letters. On the first pillar on the north side, low down, is cut, in small letters, F Owen Williams.

The windows of the arcade frame a view of the railway viaduct a mile downstream, a structure as massive as this bridge and yet possessing more elegance in its strength and a poise that raises one's spirits. Someone with whom I visited Williams' bridge thought that his design was meant 'to harmonise' in its long rectangular shape with that of the viaduct.

FINDHORN VIADUCT   1897   Murdoch Paterson   *NH 806 289*
                                                   on the old A9
                                                   through Tomatin
                                                   village

This splendid feat of iron bridge building which curves grandly and smoothly over the valley can be glimpsed from many roads in the area, but perhaps the best view is to be had from a grassy rise directly opposite it on the south side of Tomatin. It is another of the bridges built when the railway was finally brought over Slochd to Inverness.

You will see that the eight tapered stone piers get shorter as they rise up the far bank of this wide valley. On each side are two masonry spans built with the attention to detail and harmonising well with the landscape and with the main part of the bridge which is a double Warren truss topped by a crisp smaller lattice railing. It is 445 yards long with a height of 145 feet 6 inches and a curvature of 763 yards. Six of the piers go into the ground with no footwork, but the two on the river banks are set on large plinths. The bridge connects, with an embankment, to Tomatin Viaduct a short distance away. The two bridges and the embankment, which has a barrel vaulted underbridge, should be seen as one enormous undertaking.

If you go down the side road and pass under the viaduct you cannot fail to be impressed by the power of the masonry holding the curving truss in the air. Beyond is the modern road bridge on its tall double T-shaped piers, and then a small iron bridge. This was not in good shape when Bruce's team of roadbuilders reached it on the Raigbeg loop road. They strengthened it with the two sturdy timber trestles you see in place of the single flimsy one that the 1920s photos show. These three bridges make an interesting contrast and they all fit into a landscape so large it rejects little.

As you approach Tomatin village you see the TOMATIN VIADUCT (*NH 803 291*), also built in 1897 by Murdoch Paterson, but in a very different style. I have the impression Mr Paterson rather enjoyed the challenge of bringing his railway across first the Findhorn and then this dry valley in such a spectacular way. There are nine stone arches with narrow, slightly splay-footed piers and brick vaults. The maximum height of these spans is 88 feet 8 inches and they vary in width between 33 feet and 36 feet. Lime deposits have whitened the stone in places, which only serves to underline the good looks of this lofty bridge. Whereas the Findhorn Viaduct frames the entire view and so emphasises its width, depth and splendour, these arches, with their round heads, segment the land behind, each one isolating a separate picture of field and trees, a barn and cottages, hill slopes and sky. The symmetry of the arch, and

32   Findhorn Viaduct curving to the south

the lack of it in the sections of countryside behind, makes a contrast that is satisfying and exciting.

It may be convenient, at this point, to look at three very varied bridges along the line of the old A9 in this district.

The SLOCHD VIADUCT (*NH 847 238*) over the Allt du Aonaich was also built in 1897 by Murdoch Paterson when the Highland Railway was brought to Inverness by this shorter, higher route. To find the bridge, take the old road, signposted Slochd, and park at the Nordic Ski School. Follow a zig-zag mossy path through a gate, past a derelict car, and into a wood which clothes the valley. The viaduct has eight spans and the seven piers, built of bull-faced rubble, rise up the gully. The vaults are brick, 400 feet long, it is a sombre but impressive bridge built at the high altitude of 1,148 feet and normally buffeted by a keen wind.

MOY BRIDGE over the Funlack burn is down the side road to Ruthven (*NH 797 320*) and belongs in a pre-industrial world. In fact, it would be hard to find a more typical small, hill-country bridge with its low arch, slight hump, and shallow, pebbled river. It is set against a view not remarkable but epitomising this sort of country—a croft near at hand, another on the hillside, the wet track to it glinting in the sun, sheep grazing, hills and forest slopes. The soft music of the river is cut across by the lamenting cry of the curlew.

There is a mention in a newspaper of 1832 of the rebuilding of a bridge here after flood damage, so this is probably its date. It is built of

33   The timber octagons of Altnaslanach Rail Bridge

34   Detail of Altnaslanach

roughly tooled stone, a single span of 11 feet. It crosses the Funlack just after it has been joined by the Allt a Chuil, the one stream clear, the other muddy. As they flow under the arch, the two distinct colours show.

In the past there was a mill here and a larger community to serve. The road ends at Ruthven and a track goes over a wild stretch of country, beside the Findhorn, to rejoin the road at Carnoch.

AULTNASLANACH RAIL	1897 Murdoch Paterson	*NH 760 349*
BRIDGE over Dalriach		Near Moy
Burn		on the B9154

Sideways on, as you drive past this bridge, it resembles a piece of wooden Meccano placed against the hills beyond Moy Moss. The dark beams, somewhat dwarfed by their surroundings, are easy to miss, but, close to, you become aware of their size and the power of the diagonals.

It is a wooden truss bridge which takes the single track railway over the burn in a boggy stretch of land. It is 147 feet long and has five spans of 8 yard width. Each is composed of four ribs built to form an octagon, although the two end ones are half-buried in the embankments and so their shape cannot be completely seen. The ends of the large horizontal beams are rounded off. Metal plates and bolts enhance the looks of the bridge, seeming to underline its power and durability. Paterson built a timber structure here because the ground would not permit the erection of stone piers. The rushes beside the water are often laid flat, showing how swiftly the burn can run. If you go along the bank and stand right under the bridge, you will see the fine patterns which the octagons make, with beam intersecting beam. It is a strange sensation to stand with feet in the wet, observing some red and black beehives someone keeps there, listening to a train rumble overhead.

From the A9 road there is a more distant view of Aultnaslanach, one reminiscent of shots from wild west films. In America In the last century such wooden truss railway bridges were common, and many were a great deal longer and higher than this one. The pioneer railroad builders lacked stone and also masons, whereas they had plenty of timber. In this country such bridges are unusual.

After this diversion, return to the Findhorn valley bridges again.

DULSIE BRIDGE	1755 Caulfeild	*NH 931 414*
		on the Dunearn
		road, off B9007

Without a string course or a parapet to give it definition, and with spandrels and wing walls running into each other without a break,

Dulsie Bridge is a single, simple shape placed in a scene of woods, fields and water. From the grassy path which allows you to view it, you will see that the gorge it spans is not so much cliffs as knobbed ridges running down steeply to the constricted river.

The bridge is built of Ardclach granite, which is a red porphyrite. It has a semi-circular span of 46 feet, with a rise of 21 feet; but what makes it chiefly remarkable is the rake of the road, and the height above the water of 40 feet to the keystone. On the north side is a small flood arch high on the bank but not high enough to save it from complete immersion during the great flood of 1829 when the waters came within 3 feet of the keystone.

Caulfeild, building his road from Grantown to Fort George, reported that the river could be crossed at various points, all difficult and fairly costly, but he himself favoured keeping east of the river down to the coast and crossing by ferry at Forres. However he was opposed. He wrote, 'The gentlemen who lie on each side give their opinion for the situation of the bridge and the line of the road as their several reasons induce them,'—which has a curiously modern sound. Construction at Dulsie was the cheapest option, costing £150. Three companies of Lord Robert Manner's Regiment were responsible for the work. Their road continues west across open moor with long views seaward to the Black Isle and Tarbert Ness and it runs straight like a Roman military road.

To go to Ferness Bridge from Dulsie take the back road via Ardclach which is a perfect country road with mossy grass verges and woods growing close.

FERNESS BRIDGE	1814–17 Telford	NH 960 463
		on A939 north of
		Ferness

On the north side the road descends steeply to the river here, but on the south there is space for a shelf of fields and the river flows smoothly in a wide bed overhung by alders. Telford called this place Fairness which means Alder Rapids. His bridge presents a contrast to the single high stone arches at Dulsie and Daltulich above and below it. It is a neat, sturdy, compact rubble bridge with three spans and triangular cut-waters built high. They and the vaults are dressed stone. It is a slightly humped bridge still. The spans are 36 feet, 55 feet, 30 feet. If you can get down beside the river you will see the central span is not only the largest but also the shallowest. The southside arch, springing from the same level on the pier is forced into a steeper curve. The much smaller north side arch has become silted up by the bank over the years and its abutment is hidden. Telford built the spandrel walls hollow, with cross-ways

35   The soffits of Ferness Bridge

strengthening bars, and, by thus reducing the weight, he was able to have narrower piers.

During the famous 1829 floods, water filled the arches and rose 27 feet above normal. It was only the parapet ends near the banks, protruding through the torrent, that revealed the presence of the bridge. An enormous ash tree with three trunks which grew by the bridge was dislodged and sucked into the central arch where it stuck for a time and finally emerged on the far side 'shorn of its mighty honours', as Lauder describes.

DALTULICH BRIDGE (*NH 986 488*) is found by turning off the B9007 at Relugas. The road bends steeply down overhung with beeches. It is a segmental arch of coursed rubble, with a slight hump and a span of 71 feet, built on a slope. The abutments are on rock and the wing walls run round into the roadside walls. The stone used is Moinian gneiss which has touches of sandy red about it. The narrow voussoirs, like thin slats, each 3 feet long add to the looks of this handsome bridge. It was built in 1794 but the engineer is not known. It replaces one built here at the expense of a Miss Brodie of Lethen because the river crossing had claimed many lives. Her bridge only lasted a month. The present one must have been much stronger as it withstood the 1829 flood when the water rose 31 feet above normal, almost to the keystone. It is near the famous Randolph's Leap. At one time the Leap had a wooden bridge,

36   Graceful Daltulich Bridge

constantly washed away and constantly replaced by the Earl of Moray. After Daltulich Bridge was built the wooden bridge was not needed.

Standing on the bridge the hump appears off centre. This is due to the slope of the carriageway. Downstream the view, with the river bending away and a grassy slope rising, has an old two-doored barn as a focal point. Upstream there is a cliff, coarse sand beaches and a bank of pebbles. In late October small auburn leaves were drifting along the river's surface caught in thin scarves of foam; there was a soft wind and the shining of the sun on the water was almost too bright, a silver flash.

3   River Lossie in Elgin

BOW BRIG	1635 and 1787	unknown	*NJ 204 632*
BREWERY BRIDGE	1798	unknown	*NJ 223 631*
SHERIFFMILL	1803	unknown	*NJ 201 629*
PALMERS CROSS	1815 and later	unknown	*NJ 201 619*
MARYHILL BRIDGES	1867 and 1870	A Reid	*NJ 208 627*
BISHOPMILL	1873	J Willet	*NJ 216 637*

Elgin is an ancient town with much to interest the visitor, most notably the fine cathedral ruins. There are numerous bridges within a short distance of each other as the River Lossie winds through the town. If you park in the centre, you can easily walk to most of them.

37   Brewery Bridge, near Elgin Cathedral ruins

Near Old Mills is the oldest remaining bridge, called Bow Brig. It straddles old and new Elgin, near the eighteenth century mills and gives access to a modern housing estate. It is a simple bridge, without ornament, of dressed stone. The span is 47 feet and the rise 22 feet. An unusual feature is the widely splayed entrances to accommodate waiting traffic. Another detail to notice is the way the coping stones of the parapet are shaped with curved interlocking edges to prevent one being pushed off. It was the first bridge over the Lossie. On the down-stream face is inscribed Foundit 1630, Finishit 1635, but it was partly remodelled in 1787. However the side walls are original.

Brewery Bridge, which took the old road to Lossiemouth, is in the lee of the cathedral, a pretty sandstone bridge with a slight hump and two elliptical spans. There is a blind oculi over the central pier. The splay-ended parapets are unusually finished with low, flat-topped columns. The string course, accentuating the arch curve, adds to the quiet charm of this sandstone bridge. It is much used by local people who walk or cycle through the park.

Sheriffmill is somewhat similar though in an inferior state of repair. It is built of coursed sandstone rubble with two segmental arches low to the water, dressed stone voussoirs and coping, and an oculi over the pier. The feeling here is of something almost forgotten between a caravan site and the main road, overlooked by a large house in quite another style.

Palmers Cross is a bridge originally built in a similar style to Brewery and Sheriffmill (and possible all three are by the same engineer), but it has been repaired and widened so that the two faces are now unlike. The downstream side is concrete with some moulding to accentuate the rise of the two arches and the pier is a narrow rectangle with fancy semi-circular cutwaters.

At Maryhill House a small stone bridge crosses the footpath to the river. It was made to carry a path from one part of the grounds to another. Until recently it had a concrete balustrade which must have been a later addition and has now been removed. It was designed by Alexander Reid who was doing alterations to the house. The arch is semicircular but set on a skew, with the path below running downhill which presented the mason with problems, and it is interesting to see how he solved them. For instance, on the road side between the side wall and the arch there is a right-angled half pillar on one side and merely a moulded stone bracket on the other. On the riverside these features are reversed. The wing walls curve neatly up to a pillar with a coped top. The voussoir stones are dressed in a very rustic style, almost like crazy scribbling, but the edge is tooled in thin lines.

Over the river below there is a charming small footbridge with an unusual lenticular form. The upper span, carrying the deck, is made of three girder sections and the lower one is four rods. The railings are diagonal cross bars with ornamental bosses (somewhat worn now), but the graceful lines are obscured by modern mesh panels, some broken. The span is approximately 56 feet and the deck is 43 inches wide. There are steps of green slate and they have protective iron work beside them. A somewhat worn plaque says that a banker in Elgin called Robert Brander financed this bridge. The engineer is not stated but was probably Alexander Reid.

Bishopmill Bridge carried the road to Lossiemouth until by-passed comparatively recently. It is painted red, an iron bridge with large I-shaped girders spanning 80 feet. These are set into stone abutments and supported on a pair of cast iron columns. The column piers are tied by a cross girder to three small iron girders under the vaulted deck. These were almost certainly added at a later date to take the increased load the bridge had to carry before being taken out of use. There are similar supports and ties close to the abutments, which are of dressed stone, substantial and handsome with short pedimented pillars. Stamped on the iron work under the roadway is the name Fire Steel Company.

It is a functional, businesslike bridge and yet John Willet has allowed himself a few decorative touches. There are the rather pretty cast iron railings with a pattern of long rectangles and circles, and the pillar tops have neat rings round them. A lamp standard is set in the middle.

38   The skewed arch of Maryhill House Bridge, in Elgin

Unfortunately it no longer has a light. Instead there is a concrete post on a pillar top in no way in keeping with the design.

Although a product of the techniques introduced with the Industrial Revolution and probably not intended to be anything but a strong, useful road bridge, it has great appeal particularly here in Elgin where there are numerous stone arch bridges. The reflection of the railing in the water is pleasing.

V

The Upper Spey

Map 5

GARVAMORE BRIDGE   1731   General Wade   *NN 522 947*
take the minor
road from Laggan
beside the Spey

It is fitting that the earliest remaining bridge over the River Spey is also the first substantial one on the river's run from Loch Spey to the sea. Garvamore Bridge was built in 1731 by General Wade's team of soldiers and was part of the vital link from Ruthven barracks across to Fort Augustus and the Loch Ness-side road. It was typical of Wade's approach to roadmaking that he chose to drive it along the shortest route, regardless of the height of the land or its ruggedness. For those with the stamina to cross from here to Fort Augustus on the Correyairack Pass, it is a marvellous walk, and gives you a chance to see not only the well-preserved traverses by which Wade climbed up and then made the descent on the far side, where the road falls 1,500 feet in 5 miles, but also to enjoy a wild and romantic section of country which culminates in a long view down Loch Ness. (The way we achieved it, was to have two cars and exchange keys at the mid point.)

The Spey at Garvamore runs over up-thrust ledges of rock which Wade used as a footing for the abutments and the pier which is very wide, more of a central wall than a pier. This gives the bridge its particular look, and also makes it very much part of the stony landscape. Wade called it St George's Bridge, and was proud of it, being his first double span bridge. The overall length is 180 feet and the two semi-circular spans measure approximately 30 feet and 27 feet. It is humped, but the effect of the elongated pier is to give the parapet a flattened top. This parapet is 38–40 inches at the centre but tails down into the ground at the end of the long upward rise, and it has a slight splay. The road is 10 feet wide; the usual military width was 16 feet. Some years ago the

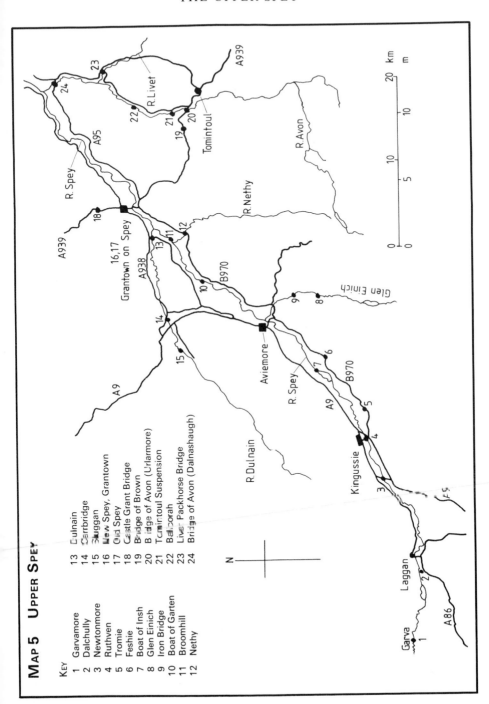

MAP 5 UPPER SPEY

KEY

1	Garvamore	13	Dulnain
2	Dalchully	14	Carrbridge
3	Newtonmore	15	Duggan
4	Ruthven	16	New Spey, Grantown
5	Tromie	17	Old Spey
6	Feshie	18	Castle Grant Bridge
7	Boat of Insh	19	Bridge of Brown
8	Glen Einich	20	Bridge of Avon (Urlarmore)
9	Iron Bridge	21	Tomintoul Suspension
10	Boat of Garten	22	Balcorah
11	Broomhill	23	Liver Packhorse Bridge
12	Nethy	24	Bridge of Avon (Dalnashaugh)

39   Garvamore Bridge. Note the buttresses and the staying rods

bridge was falling into ruin but an excellent repair job was done by a team of volunteers.

The beauty of the river and the grandeur of the bare hills impress whatever the weather; but if you visit it, as I did, on a warm day in August when the bracken is yellowing in patches on the mountain slopes and the light strikes sparks from the rock-scarred summits, and the quiet peace is underscored by the continuous sound of water and the occasional cry of bird and sheep, you may well linger sometime.

The present solitude of this powerful bridge, striding the river in two paces, holds within it a different moment of history when the soldiers and masons with their attendant suppliers were camped here, no doubt with much mud, swearing, exhaustion, hunger and thirst. At the hamlet of Garvamore, half a mile downstream, the ruined two-storey building with the temporary roof is the original Kinghouse. It was used as a barracks-cum-inn by the troops (or their officers at least) and by the travellers who used the bridge and the Corrieyairack road when it was built which, amazingly, only took the one summer of 1731. The men were supplied with shovels, picks, spades and crowbars. Obviously gunpowder was used, and doubtless barrows, carts and iron hammers. When the job was done Wade gave a feast, roasting six oxen for the six companies involved. He called his roadmaking troops his Highwaymen and seems to have been unusually thoughtful for their well-being. As well as these occasional feasts, he arranged for the excise duty on the

40 A bridge no more: Glen Shirra Bridge left behind in a field

beer they brewed themselves to be set aside so that they could quench their thirst without penalty. Previously the excisemen had made use of the very roads the soldiers were constructing to approach and demand the tax. The troops in the road gangs had double pay, that is one shilling a day rather than sixpence. For this the rate of progress was expected to be $1\frac{1}{2}$ yards per man per day, working a ten hour day. The captains in charge of each 100 men, two subalterns, four NCOs and a drummer had no extra money as they did not actually bend their backs. The subalterns were in charge of having huts put up to shelter the men and of 'making provision for their tables'. Wade had as many as six hundred men working for him in mid season. Later under Caulfeild, the numbers were greater, the time each man worked less. If the weather was so bad that no work was done then they did not get double pay. Because of the inclement winters roadbuilding was restricted to 1 April to 31 October, but in the midsummer months more men were drafted to the roads than in spring and autumn. Anyone who has experienced the exceeding chill of many April days, with snow showers and horizontally driving winds, will sympathise with the men who toiled on this bridge.

Just beyond the old Kinghouse the road crosses a canal connected to the reservoir and then you pass the small Glen Shirra Wade bridge in a field, left high and dry by both river and road (*NN 554 932*).

At Laggan there was once a Telford bridge which was replaced in 1828 by a timber truss, then by a steel bow in 1913. All these are now gone but an interesting plaque on the old abutments shows the profiles of them. The modern bridge is not exciting.

Near here is another small Wade bridge. Go south from Laggan on the A86. About a mile after the left turn to Dalwhinnie, as you climb the hill, look out for a farm road immediately before a small fenced plantation of young conifers on the right. (The plantation on the left is older, larger and not fenced.) This lane will take you, in a quarter mile, to the little bridge over the Mashie Burn called DALCHULLY (*NN 600 931*). Built in 1731, it is part of the same military road which had come from Catlodge. It joined the present road to Garvamore at a sharp bend just before Spey Dam. Dalchully has one semicircular span of 22 feet and extensive spandrels and wing walls. It is rubble, barely coursed, and humped. It is approximately 12 feet from the water to the crown of the arch. The splay of the parapets and a decided curve to the spandrels means that the hump is the narrowest point. The Mashie is reedy and flows quietly in gentle curves to the Spey.

Some way farther along the A86 towards Spean Bridge, you pass the gatehouse lodge to Ardverikie House, built in 1878 by John Rhind. It is an elaborate little building such as Victorians liked to put at their gates and beside it is a bridge with a cast iron, diamond lattice railing and stone abutments. It is entered through cast iron gates hung on tall stone posts. Whether this is also Rhind's work I cannot discover, but it seems likely. It crosses the River Pattack flowing out of Loch Laggan past its beautiful terraced sandbanks.

NEWTONMORE BRIDGE	1926	Owen Williams and	*NN 709 980*
		Maxwell Aryton	on the B9150 near
			Newtonmore

This handsome bridge, built when the old Parliamentary road north was modernised and became the A9, was the largest one. Now off the run of the new A9 it is well worth turning aside for. It replaces a fine Parliamentary structure that had three main spans and two flood arches on the north bank. It is a pity that Major Bruce did not merely by-pass the old bridge; but perhaps the lie of the land did not allow for this.

It inclines from south to north across the 260 feet wide valley with the north approach road embanked. The three arches decrease not only in height but also in width, the spans being 107 feet, 87 feet, 67 feet. The voussoirs and parapet are moulded and there are semi-hexagonal refuges at each end.

Unlike Williams' bridge at Findhorn, this is the traditional shape for a masonry bridge and yet the concrete allows for effects that coursed

41   Newtonmore, a fine bridge neglected

42   Detail of the jutting keel-like cutwaters on Newtonmore Bridge

stone cannot achieve in a bridge of this size. The spandrels are curved outwards, being 5 feet thick at the bases and only 1 foot at the parapet. The designers used the vertical expansion joints in the concrete to articulate this spandrel curve, but this is less evident now than they intended because of the growth of a dark lichen. The shape of the cut-waters is another feature that underlines the forward and outward movement of the arches. The smooth voussoirs come round and jut like the keels of boats, or massive arrowheads. The piers have stepped bases and are set on rocky outcrops in the river which here is gently flowing with pebbly banks. It has a normal maximum depth of 30 feet. Only the largest arch is normally over the river, the smallest spans a field where a lorry is now parked and rusting, hinting at the comparative neglect of this fine bridge which has taken a traditional shape and given it new and exciting life.

The shape is reminiscent of some much smaller bridges on the A9 also designed by Williams. A typical one at Crubenmore is described on page 123.

There are splendid views here of the surrounding hills, and you can see a two span, iron truss railway bridge crossing the Spey just down-stream, its pier on a small wooded island.

From here it is worth a short detour from the Spey to see two stone bridges over tributary rivers.

Find the road to Tromie Bridge by turning in the centre of Kingussie and crossing the Spey by RUTHVEN BRIDGE (*NN 759 997*) which was built in 1894. It is a three span, iron truss bridge with stone abutments and oval stone piers. Just after it you pass Ruthven Barracks, the node of the military roads in this area and still an impressive and even menacing building. TROMIE BRIDGE (*NM 789 995*), on the other hand, is picturesque and a delightful place to stop because it spans the River Tromie at a particularly pretty stretch of the fierce little river. Wade built a bridge here, but this is an early-nineteenth-century replacement, which has itself been much repaired. It is built of rough rubble with a slight east–west slope. The segmental arch is almost a semicircle and has a span of 28 feet. It is a classic example of abutments built on the rocky sides of the river, tumbling down through lichened and mossy stones and banks where wild flowers grow.

Continuing north up the B970 you pass a magnificent stand of pines growing on a slope with thin heather beneath and a drystone wall beside the road. After approximately 4 miles you come to FESHIE BRIDGE (*NH 852 043*).

The River Feshie pouring down from the mountains is restricted here by a sloping bed of stone which juts out into midstream and forces the water into a narrow race where it foams and becomes the colour of mint

43   Feshie Bridge, footed on rock

icecream. This makes an obvious place for a bridge as the rock outcrops make an excellent foundation for the abutments. Feshie Bridge has one main span of 40 feet and a small flood arch of 12 feet with a large spandrel wall between, and it has been slightly angled to fit the rock platform. It, like Tromie, is an early nineteenth century bridge that has been stayed with rods through to strengthen it and repaired several times. One iron girder on the east bank is prominent. The bridge is built of the simplest rubble materials and without decoration. It perfectly suits the wild hilly place where it is set, with alders crowding to the water's edge and some fine conifers downstream. Below, the river widens to a milky pool, and upstream the grey rocks have been cut into shapes so smooth and rounded they look more like moulded grey plastic for some strange piece of equipment than stone.

From here it is a short drive back to Kincraig on the A9. Where the River Spey leaves Loch Insh it is crossed by BOAT OF INSH BRIDGE (*NH 835 055*). This is a late-nineteenth-century ten span wooden bridge. It has trestle piers faced with iron and two stone piers which are second and third from the west and east banks respectively. Metal girders carry the deck and the railing is of wood with sheep netting. The west abutments have curved wing walls and low rectangular pillars, one without its coping stone; and on the east bank there is an unsightly tumble of rocks. Altogether the bridge is not in spanking repair and its looks are further marred by the three pipes running across it at deck level on the upstream side, the largest encased in black plastic. It has nothing of Broomhill's crisp detailing and handsome triangularity. Those who would like to see a splendid timber bridge should not fail to see Broomhill (page 68).

However, it is worth stopping at Boat of Insh (the name implies there was once a ferry here) to see Loch Insh, which must be one of the finest small lochs. From the bridge you look south up the Spey valley. On your left is the white church on its promontory and ahead is an island so loaded with trees it is like a boat crowded with green sails. Islands of emerald grasses and rushes sprout in the grey-blue water and, when I was there in July, a family of swans was serenely floating in their protection. In the distance yellow and blue sailed boats were passing, and near at hand swallows skimmed under the bridge. Over all rise the harsh precipices of the Cairngorms.

At Aviemore I again suggest a diversion from the main river to see two of the mountain bridges erected for the convenience of walkers and climbers.

GLEN EINICH Mountain Bridge   *NN 924 043*
over Am Beanaidh   walk from Loch an Eilein

The walk to this bridge and back is about 8 miles along a well-made path without steep gradients. It is worth doing for itself as Glen Einich is a most beautiful glen. The path is parallel to Am Beanaidh for some of the distance and rises and falls, giving varied views of the glossy pines by the water and the massive flanks of Braeriach which rises to a plateau of 4,000 feet. You park at Loch An Eilein (or at Coylumbridge) and follow the signs to Glen Einich. The bridge is about a mile from where the trees by the riverside end and the path descends to the water's edge. You will see several ice-cut high ramparts ahead. The bridge is under one.

It does not look particularly strong, the deck slats tip sideways at certain points, but it must have considerable sturdiness to stand up to the river in spate and carrying lumps of ice and rolling stones. It is constructed from three pairs of tree trunks, long strong ones, the largest diameter being about 9 inches. These are set from the abutments to two low narrow rectangular piers standing on stones in the water and buttressed loosely with boulders. The piers are of coarse concrete and have beams bolted to their sides and tops. The tree trunks, or poles, do not appear to be bolted to these beams, but lie on them and are held in place by wires, of no great thickness, and wooden wedges. Each pair overlaps the next pair and slots between it. Wood slats, measuring 9 inches, are nailed to the poles to make a walkway. There is a two-strand wire rail on the upstream side up to a height of 26 inches. The whole span is approximately 19 yards.

44   Glen Einich mountain bridge

While we were there the bridge was much used by parties of walkers and by a character with a bike. The way on to the loch is rewarding for those with the legs and footwear for it.

On your return, turn right at the junctions of paths along a track marked Lairig Ghru. After about half a mile you come to another mountain bridge called IRON BRIDGE (*NH 927 078*). This is more substantial and to that extent less romantic. Set wide and high over what can be a fierce river, it was built in 1912 by the Cairngorm Club of Aberdeen for the use of walkers, especially those doing the Lairig Ghru walk to Braemar. On the bridge is a metal plate giving details of this famous 24 mile walk.

Iron Bridge is a single girder span of 47 feet set on concrete abutments. The girders measure 15 inches by 5 inches and are at an approximate height of 7 feet from the water. The deck is of metal plates 4 feet

45   Broomhill Bridge, a timber truss bridge

wide with a firm 42 inch high railing. The steps up on the left bank are
of perforated metal; on the right what was probably a metal ramp has
gone.

Now pick up the River Spey at Boat of Garten. A ferry once ran here.
It was a chain ferry and an old photograph of it, and of the wooden
bridge that replaced it, can be seen in the hotel. The wooden bridge is
now itself replaced, but something similar can be seen at Broomhill.

For those interested in railways and steam, Boat of Garten is a good
place to visit as it is the headquarters of the revitalised Speyside Railway
and has a small museum. The cast iron bridge over the line was brought
here from Dalnaspidal and replaced an identical one; both were cast in
Inverness at the Rose Street Foundry. It is very like the one at Garve,
except that this one retains the original lamps which are much more in
keeping.

| **BROOMHILL** | 1894 | John MacKenzie of Kingussie, Charles MacKay | *NH 997 223* off the A95 towards Nethy |

From the A95, where it runs half-way up the hillside, you have a good,
long view of a remarkable timber bridge lying across a broad sweep of
silver river in a flat valley—but you need a sunny day. In overcast
weather the bridge can be swallowed into the landscape, its timbers
merging with trees and gravel.

46   Detail of Broomhill

It was built in 1894 at a place where other bridges had been swept away. One built in 1857 was probably the first. Its construction was chiefly due to the minister's 'zealous and unwearied advocacy', as the *Inverness Advertiser* reported. He hoped that a bridge would bring the scattered communities together.

The present Broomhill Bridge has fifteen spans and the piers are timber trestles. Nine of these stand in the water, with three on the south bank and two on the north. These trestles are built of five posts, the three central ones being perpendicular and the outer two set at an angle and faced with metal. Horizontal planking comes up half their height, and the top section is braced with diagonal members, one on each side. On top is a massive beam with rounded ends, and on this, set transversely, are five beams supporting the road timbers. The five piers over the main river channel have triangular trusses above. These are the most noticeable features of the bridge, and are 20 feet wide at the road and 6 feet 3 inches tall. The side beams continue below deck level and provide ties for the trestle beams. They are riveted with metal plates and have three railing bars to a height of 40 inches. The abutments are tooled granite. There is almost no metal in this bridge which was partly reconstructed in 1987.

There is a pleasant grassy path by the river from which this handsome bridge can be inspected and all the interlocking timbers clearly seen. It is a fine sight, with its gold-brown triangles standing against the water,

47   A bridge no more: Railway underbridge near Dulnain

and particularly so in being one of the last such timber bridges to remain.

From Broomhill is a short drive to NETHY BRIDGE (*NJ 002 205*) over the Nethy. It is one of three Parliamentary bridges on the B970 near here—Duach, Nethy and Aultmore. The Nethy Bridge was partly destroyed in the 1829 flood. Lauder describes a sawmill being moved bodily off the bank and coming towards the bridge 'steadily and magnificently like some 3-decker leaving dock.' The mesmerised onlookers watched it come towards the bridge but at 100 yards it hit something and fell to pieces. 'Spreading itself all over the waters, it rushed down to the Spey in one sea of a wreck.' The bridge lost one span. The present bridge is a re-working of Telford's. His spans were 24, 36, 24 feet and this has three unequal spans, the most northerly now blocked off. The splayed parapets have unusual square ends with pyramidical caps. This is an unostentatious, country bridge in a quiet village.

From Broomhill, in the opposite direction, you come to DULNAIN BRIDGE (*NH 997 249*) at the junction of the A95 and the A939.

A bridge built here in 1791 was destroyed in the 1829 flood and Joseph Mitchell built this replacement in 1830. It has a single span of 65 feet. It is a handsome bridge of tooled granite with ashlar voussoirs which are hard to see because of the way it has been widened. The parapet was removed and rebuilt on a widened deck which is supported on cast iron joists. The almost continuous traffic of heavy lorries

48   Sluggan Bridge. The arch ring remains

passing over the bridge in summer makes one wonder how long it can endure, considering Mitchell designed it for horses, carts and carriages.

The River Dulnain has two other bridges of great interest, so that a detour here is recommended.

At Carrbridge you could once have seen a fine elegant concrete bridge designed by Owen Williams to complement the ancient bridge here, but it has sadly been demolished. You can however enjoy CARR BRIDGE (*NH 906 229*) in all its antique charm. There is an informative plaque on the site which will tell you most of what is known of this bridge. It is known as a funeral bridge because it was chiefly needed for the conveyance of coffins to the burial ground. A mason called John Niccelsone built it and the parish paid for it. It cost £100—quite a considerable sum in 1717—and when it was built was called Lyne of Dalrachney bridge. It is 2½ yards wide, but between parapets this would have narrowed to 7 feet only. So it was clearly only meant for men and horses. It is a high hump bridge set on the rocks of the river side, and somewhat dwarfed now by modern buildings and street furniture. Very little remains apart from the arch ring, showing its inherent strength.

SLUGGAN BRIDGE   post 1829   *NH 870 220*
take station road in
Carrbridge and continue

Approximately 1½ miles from the station at Carrbridge the road rises at a place where you can pull in on the left beside a high stile. You can see the straight line of the old military road coming up from Kinveachy. You must take the right hand track, through a chained but broken gate, and follow this road downhill. The bridge is at the most 15 minutes walk. You can see it, with its grassed roadway, from the hilltop but lose sight of it as you drop down into the flat valley bottom. This can be wet.

The bridge appears suddenly beyond an abandoned farm and half-hidden by bushes. After the toylike view from above, its height surprises, the semicircular curve rising above the shallow brown river. The lack of parapet emphasises the beauty of the arch. It seems the essence of a bridge.

The precise dating is difficult to come at. It is not Wade's bridge, though on the line of his road, and has probably been rebuilt and patched several times. One authority says it was rebuilt after the 1829 floods, another that it was not damaged by them, and a third that it was abandoned in the period 1798–1813 when the road was realigned by the Parliamentary Commissioners. This seems to be most likely.

The bridge has a span of $19\frac{1}{2}$ yards and a rise of half that. It is approximately 14 feet wide now without its parapets. Built of coursed rubble with thin slab voussoirs, it has long wing walls which on the north bank join the spandrels with a clearly visible join. The ground on both sides is flat, although there is a wooded cliff just upstream, and the height of the arch is doubtless to cope with flood water, draining from the Monadhliaths. An old birdcherry tree grows against the south abutment, that delicate northern tree with upstanding white flower spikes, more like a wild lilac than a cherry. When I was at Sluggan in early May blue-caverned summer rain clouds piled over the green hilltops, the cuckoo was singing, wagtails were springing into the air with flirts of their tails over the pebble beds, a redstart flew out of the birdcherry, and the swallows skimmed continously along the river, under the arch and in and out of the barn doorways.

You can cross the bridge and follow the military road or the many other paths. There is a charming, almost fairytale feeling about this old arch with its grass road and rickety gates, the deserted farm and the wide, lonely countryside peopled with sheep, rabbits and birds. How very different it must have been when it was the main road north.

NEW SPEY BRIDGE   1931   Blyth and Blyth   *NJ 034 268*

on the A95 at Grantown

If you walk over this bridge you will probably be struck by its solidity and spaciousness. There is very little in the way of decoration—just finely inscribed lines on the parapet walls and panelled inner sides to the

49   Spey Bridge at Grantown. Note the four ribs

large five-sided refuges, which offer ample space, as do the pavements and the roadway. In the centre the old county boundary is marked.

You must go along one of the riverside paths to appreciate the powerful grace of the arch which soars over the river. It is a segmental span of 240 feet with four gigantic cross-linked ribs clearly visible. There are semicircular flood arches on the banks. The abutment walls which become the polygonal refuges, flare out at ground level into stone skirts which act as cutwaters in the event of floods. It is a boldly simple shape that suits perfectly the single grand gesture of the plain arch. The wide cuff of the voussoir rim is the only external decoration, apart from the entablature with the date.

For me this bridge is a far more convincing use of concrete than the fussy narrow bridge at Culdrein (page 82) or the heavy, almost brutal one over the Findhorn (page 46). It makes an instructive contrast with Caulfeild's stone arches a short distance downstream.

OLD SPEY BRIDGE	1754	Caulfeild	*NJ 040 263*
			left bank riverside
			road off A95

This fine old bridge was one of the most substantial built by Caulfeild when building the military road from Braemar to Fort George via the Lecht and Grantown. It is not much altered, though much stayed, and the flood arch, damaged in 1829, had to be rebuilt.

There are three unequal spans of 25, 45, 80 feet and the deck is on a slope. This has been necessitated by the lie of the land, for there is a steep rock cliff on the south side but alluvial banks on the north. The river has sandbanks and tiny islets of grass. The spans are linked by wide spandrels which some have seen as a design fault, but they certainly are part of the bridge's charm. The triangular cutwaters are extended up into half-hexagonal refuges, one pair 11 feet wide and the other 15 feet. On the left bank the parapets are neatly finished with a curved edge, but on the right bank become high dry stone walls which continue beside the road for some way, overhung by beeches. From here is the handsomest view of the bridge through the trees. The water, in satiny strips of dark grey and silver, and the sycamore trunks are almost the same colour as the stone. On this side is a broken date stone saying:

D 1754

E COMPANIES

OF THE 33 RD REG

IMENT THE RIGHT

HONOURABLE LORD

CHARLES HAY

COLONEL

E N D E D

The word 'ended' (followed by a mason's mark) is written in larger letters and has a thankful look.

It is a pleasant, quiet spot, a fisherman's place. Wooden steps lead down to riverside paths. In the early evening men with rods were arriving, intent, serious, self-contained. They cast the briefest of uninterested glances at my camera but peered over the parapet keenly. Dark, quick water, brilliant with weeds; the house overlooking the bridge as still as everything else.

Not far north of Grantown on the A939, the road to Nairn passes under CASTLE GRANT railway bridge (*NJ 033 302*), which is one of Mitchell's more extravagant exercises in Highland baronial style, built in 1863.

It is a skew bridge and asymmetrical in design. On the west side the abutment curves down in a wing wall that finishes with a low pillar, whereas on the east the wall continues at a level height bending round to join the tower of the gatehouse. As the embankment for the line needed more support than this, there is a further retaining wall, invisible from the road, which also abuts the tower. It is built of tooled granite and the span is 48 feet. There is a castellated parapet and half-turrets with arrow slits. Altogether, it reminds me powerfully of a stage set and I half expect the operetta chorus to arrive with tambourines.

Lord Seafield who owned Castle Grant in 1863 did not want trains in his parkland. Mitchell attempted to find a lower level for the line but failed. Lord Seafield was offered compensation which he rather grandly declined. The company then built him this 'pretty lodge', as Mitchell calls it, to thank him for his 'liberality in declining compensation'.

From Grantown it may be convenient to explore the River Avon and rejoin the Spey farther down. Take the A39 towards Tomintoul. It runs over a particularly spectacular stretch of high country with superb views of the Cairngorms to the south and the Hills of Cromdale to the north.

The **BRIDGE OF BROWN** (*NJ 124 206*) is a Caulfeild military road bridge, now by-passed. It was built in 1754 over the Burn of Brown in a steep dip in the moor. It has a low segmental arch of coursed rubble.

The **BRIDGE OF AVON** at Urlarmore (*NJ 149 202*), also a Caulfeild bridge, is rather disappointing. It was constructed in 1754 by Lord Charles Hay's Regiment as a stone at Lecht testifies. (They called the bridge Campdalmore.) It has been a beautiful bridge but is now defaced by a black bailey bridge section placed on top, after the severe winter of 1978 weakened the structure. As well as the metal work, there is plastic sheeting protruding below, traffic lights, the chug and roar of accelerating vehicles and, when I visited it in 1988, the smoke of fires as men cleared the hillside above to make a new road. Not the place or atmosphere in which to appreciate the solid dignity of an eighteenth century design. However, down at the water's edge, some of the bridge's fine points can be seen. It is a two span rubble bridge with segmental arches, one larger than the other, and triangular cutwaters. It is to be hoped that, when the new road is made and the bailey bridge removed, every effort will be made to restore the bridge and replace its parapets. In the museum at Tomintoul old photographs of the bridge can be seen.

From here take the minor road B9136 north, or, preferably, return towards Grantown less than a mile away and take a small road on the right which runs along the River Avon and is a pretty and interesting road. About 2 miles along you come to **TOMINTOUL SUSPENSION BRIDGE** (*NJ 146 231*). When I first saw this bridge it was not to be crossed safely, but it has since been repaired by the Senior Reserve Regiment of the army, as a notice tells.

The cable has a high catenary curve which alters the usual profile. The towers are right angled steel arches and the cables are bedded in concrete on the banks. There are also two cables under the deck anchored into the same places. The wooden deck is hung in right angled U-shaped metal brackets which are the same shape as the towers but

50   A remaining span of the old Packhorse Bridge over the River Livet

reversed. It is this repetition of shape that gives this bridge such a look of efficient commonsense, as befits a military construction. There are four steel wire stays from the centre banks, forming a cross.

In the next field are the remains of a former suspension bridge (*NJ 146 233*). It is clear that the B9136 has been considerably altered in level.

Several miles farther on the road crosses the Avon at BALLCORAH BRIDGE (*NJ 155 265*). It is called the Silver Bridge locally and is late nineteenth century in date. The engineer is unknown. It is an important crossing because the Avon has few bridges. The height suggests how tempestuous the waters can be. It is an iron bowed truss set on ashlar abutments and with a wooden deck. A pair of transverse girders in the centre are extended to form stabilisers. The risk with all truss bridges is of twisting and snaking in high winds. The railings extend beyond the bridge and finish in fence-like metal posts. From the waterside, again much frequented by fishermen, the lay out of girders and cross members and deck is clear and the silvery diagonals of the bridge are held against the sky.

Just beyond the junction of the B9136 with the B9008, below Glenlivet Distillery, is the charming OLD PACKHORSE BRIDGE (*NJ 198 302*). It has miraculously survived time and flood and lies on a sharp bend in the River Livet. Originally it had three spans but that on the far bank was lost in the infamous 1829 flood. It has been founded on the rocks of the river bank which are canted up. Some stick out of the water

51   Bridge of Avon, Dalnashaugh. A romantic arrangement

above the bridge and are tufted with grass. All the spandrels have long since gone and only two arch rings are left, picturesquely grassed over. One is left to imagine how the bridge would have looked with a continuous road from one side to the other and a traffic of laden pack animals, rough carts, itinerant tradesmen, the gentry on horseback, shepherds, and herded animals, women going to market, playing children and much more. The fragile-looking, thin arch rings are a good demonstration of the strength of an arch with all the separate stones in compression supporting each other.

The date of the Packhorse Bridge is unknown. It may have been connected with Blairfindy Castle which was built in 1586. It was certainly an important place to cross an unpredictable river. Now it seems forgotten, tucked behind several modern cottages, surrounded by wild roses and broom brushes. Because date, designer, history and original appearance are all more or less unknown, there is not a great deal to be said about it except—do not miss it.

You can now follow the B9008 down to Dalnashaugh, where the Avon joins the Spey. Here, from the modern bridge, you can see the old AVON BRIDGE (*NJ 184 359*) on the A95. It is of mellow, dark stone held now with metal stays, a rubbled bridge with narrow voussoirs and rounded cutwaters. The main arch is strong and graceful, curving over the brown water often flocked with foam eddies which gather in the slacker water by the smaller arch, wrinkled and pocked like the skin of old cream. It is no longer used except by walkers and has become an adjunct to Ballindalloch Castle gatehouse (1850) which is on the east bank.

The modern bridge is a supported cantilever with two huge sloping legs of a certain elegance. It is interesting to observe the three sorts of architecture displayed here—the Victorian baronial with a fairytale pinnacle, the plain eighteenth-century stone arches, and the concrete stride of our own century.

VI

The Lower Spey

Map 6

Take the cross road between the A95 at Cromdale and the B9102 to find the BOAT OF CROMDALE BRIDGE (*NJ 066 289*), a high angular bridge somewhat at odds with its rural setting and the old church beside it. It is a flat-topped truss, painted grey-green, and set on concrete abutments and a central pier of two columns. The truss is in two sections and constructed with diagonal girders 10 inches wide in pairs 2 feet apart and held by cross struts which form fourteen lower triangles and thirteen upper ones, so that the top is slightly shorter than the deck level. There are cross bars overhead and railings at each side. So many hard diagonals result in a complicated grid being placed on the background countryside, a grid that shifts as you move.

Some beautiful trees are close by. On the steep bank a group of old pines with canted trunks, a scabby russet; an ash, alders, sycamores, a big pussy-willow, a beech defiantly growing despite savage lopping and a chestnut in the church yard. Here a wall tablet commemorates a soldier from Inverallan who died at Fort Abouzaie in the Punjab in 1855. I wonder whether, as he died in that hot dusty fort, a vision of wet, green Speyside came to him.

I can find no record of the date of this bridge or of its engineer. It is almost certainly early this century. There was an earlier bridge and the abutments with their pepper-pot decoration may well belong to it. From beside the church gate you look along the bridge, as though down a metal corridor, through which the blues, greys and greens of the landscape are seen; from here it harmonises well and its strict geometry enhances the fluidity of nature.

A similar cross road, from the A95 near Advie to the B9102, will take you to CULDREIN BRIDGE built in 1922 by Alexander Hogg and the Yorkshire Hennebique Company. There is something of the mock castle about it. It is a reinforced concrete truss bridge of three spans carried by two pairs of circular piers and matching abutments. The

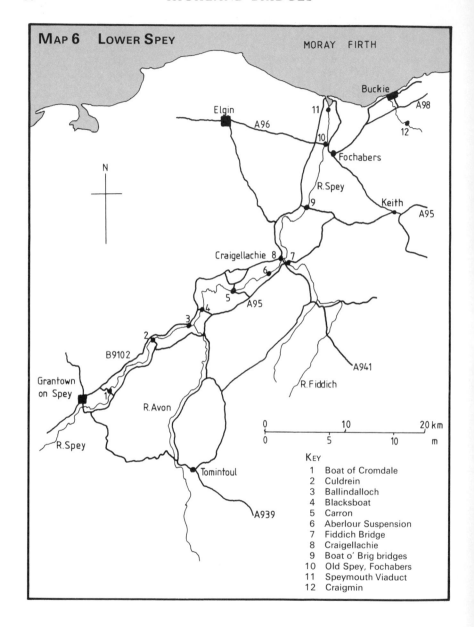

MAP 6 LOWER SPEY

MORAY FIRTH

Buckie

Elgin 11 A98

A96 12

10

Fochabers

R.Spey

9 Keith

A95

Craigellachie 8 7

6

5 A95

3

2

B9102

Grantown
on Spey 1

R.Avon A941

R.Fiddich

0		10		20 km
0	5		10	m

R.Spey

Tomintoul

A939

KEY
1   Boat of Cromdale
2   Culdrein
3   Ballindalloch
4   Blacksboat
5   Carron
6   Aberlour Suspension
7   Fiddich Bridge
8   Craigellachie
9   Boat o' Brig bridges
10 Old Spey, Fochabers
11 Speymouth Viaduct
12 Craigmin

52   Boat of Cromdale, a metal truss

trusscs are actually set on the piers on short legs tied with cross girders,
The abutments are finished off with stub pillar tops to match the pier
columns.

What makes the bridge unusual is the piercing of the truss walls with
octagonal apertures. Each octagon is set with five bars at 7, 14, 23, 35
and 48 inches height, and each one is sub-divided by a central pillar
measuring 6 by $4\frac{1}{2}$ inches. The edges of these small posts have been
chamfered. The outer apertures in each span are smaller than the
middle ones, and where the spans join there is a square pillar matching
the abutment ends. The octagons are quite large (64 inches high and 88
inches wide) yet they do not give a feeling of airiness. The bridge is heavy
and looking down it is like looking along a stone corridor, the gaps
being hardly apparent, unlike the open lattice effect of Boat of
Cromdale. Clearly much thought went into the detailing of this bridge

53   Culdrein Bridge, a concrete truss

but the result is fussy rather than harmonious. It is interesting to compare it with the huge Findhorn concrete truss bridge (see page 46), built only a little later and much more famous, but in essence not dissimilar.

In October there were many pheasant in the fields around, and a dead one in the gutter of the bridge with its scarlet eye and scalloped feather patterns splashed with mud. Rosehips were thick in the hedges and a buzzard hung in the sky where lakes of watery light lay between the clouds.

BALLINDALLOCH   1863   G McFarlane of Dundee   *NJ 168 368*
park at the old
station on B9137

This 117 year old bridge, built for the Speyside Railway, looks fresh and new with its crisp central name plate announcing its engineer. It is a wrought iron, single span, flat-topped box girder 198 feet long, with plate girder access spans. The massive stone piers on the banks take the truss 20 feet above water level. The access spans have neat cast iron railings in the 'church window' pattern.

The railway closed in 1968 and there is now a wooden walkway where the rails ran and yet there is still a feeling of transport here, perhaps because the black ruts and grass centre of the track running between the bushes look not unlike railway lines. Small signs of the past are about—

54   Ballindalloch Railway Viaduct, the engineer's signature

a rope hand-hold fixed to cast iron posts, a run of pulleys on the bridge
to take a signal cable, and the station itself, cleanly painted for all its
weed-filled track. Day excursions from Aberdeen to Speyside cost two
shillings and six pence at one time, but the line was used more by freight
trains, especially serving the distilleries. Beside the station are old sheds,
numbered cattle pens and a weighbridge. Lupins, broom and rabbits
abound now.

The main lattice span of the bridge is constructed from girders
23 inches by $7\frac{1}{2}$ inches erected in pairs to form diamonds that are
$26\frac{1}{2}$ inches square. The girders leaning towards the centre have cross
stays, those leaning outwards none. One imagines from the way his
name is inscribed over it that G McFarlane was proud of his handsome
and durable design.

Again you must take a cross road to reach the next bridge at

55   Cast iron girder railway bridge at Blacksboat. Note the cast iron plate sides

Blacksboat. Turn off the A95 on to the B9138 at Marypark, and run downhill to BLACKSBOAT BRIDGE (*NJ 185 390*). Another somewhat utilitarian bridge spanning the river near the restored station. It was built in 1911 by Alexander Findlay and Company of Motherwell and is a double Warren truss in three sections supported on two pairs of cylindrical stone columns tied with iron girders. These are exceptionally well built of tooled stone, tapered, and finished off with flat round tops like millstones. The abutments have narrow L-shaped tops, so do not match. The lattice is diagonal and rises 62 inches above the deck with curved outer stabilisers and three inner safety rails.

Between this bridge and the station are two small road bridges built by the railway. They have three cast iron girders and wooden transverse beams and the sides are cast iron panels. One crosses the old track and the other, judging by the ramp-like bank in the field below, perhaps crossed a siding. Such cast iron girder bridges are very uncommon in this area. The thin panel plates of the sides measure $28\frac{1}{2}$ inches by 51 inches and they are strengthened by thick splayed side edges. One parapet is topped with a right-angled rim, the other a rounded rim, which makes you think a repair has been done until you see the second bridge rims are the same. There is an almost identical bridge at *NJ 128 346* near Advie but this has more decorative panels.

56   Carron Bridge, showing underside of the bow truss

| **CARRON BRIDGE** | 1863 | McKinnon and Company, Aberdeen | *NJ 224 412* leave A95 at Bridge of Derrybeg, approximately 2 miles |

As one approaches this bridge it appears something of a contraption and perhaps a litle scary to cross, with only a thin metal fence between oneself and open beams over the water. It is not until you go down to the riverside and look up at the high iron arch that you appreciate this handsome bridge. (You can go down the bank with ease from the right bank where there is a convenient slot for a car.)

A bow truss built in 1863 for the Speyside line, it was the last cast iron rail bridge in Scotland. (Cast iron was superseded by wrought iron and then by steel.) It has three ribs, each cast in seven sections and a lattice infill (or spandrel) of diamonds which get progressively smaller. The truss is set at right angles to the abutments; this contrasts with Craigellachie (see page 87). There are massive rustic ashlar piers, with triangular cutwaters and semi-hexagonal towers with refuges, and matching abutments with substantial flood arches between the two. Two string courses emphasise the top and bottom levels of the iron work. The masonry is weighty, the river wide—both enhance the spring and lightness of the arch.

At road level it will be seen that the railway track side of the bridge

57　　Craigellachie Bridge

has lost its deck. This enables you to take a good look at the construction—the I-beams laid across the lattice work above the ribs. There are three sorts of fence—horizontal rails, vertical bars and metal mesh.

I visited it on a misty, windy October evening when the dark waterside pines were sombre. Duck flew up from the river which was coughing and gurgling on its stones. Dead knapweed heads on the bank clacked together.

In the town of Charleston is a pedestrian suspension bridge called ABERLOUR SUSPENSION (*NJ 263 429*). You will find it in the park. It is long, narrow and low to the water. The pylons are tapered lattice girders finished with ball and spike finials and cross braced. The walkway is a diamond lattice girder hung from the suspension cables by links. It was built about 1900 by James Abernethy and Company of Aberdeen. This stretch of the river is obviously a great spot for fishing. When I was there on a May morning the banks were dotted with silent, concentrating men.

At Craigellachie, apart from Telford's great bridge, there is a handsome early-nineteenth-century stone arch bridge spanning the Fiddich, a river which joins the Spey at this place. FIDDICH BRIDGE (*NJ 294452*) is built of coursed rubble with narrow voussoirs and a dressed stone parapet and string course. It has small buttresses on low plinths. The road rises from west to east. Beside it is a small inn, cheerily painted red and white and with two chamaecyparis trees clipped to look like green toadstools, and an old apple tree laden with fruit in autumn.

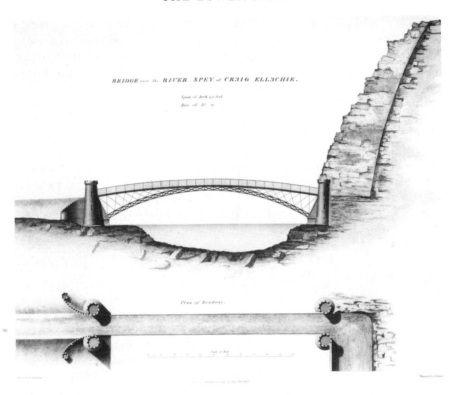

BRIDGE over the RIVER SPEY at CRAIG ELLACHIE.

58   Telford's drawings of Craigellachie, taken from the *Atlas*

CRAIGELLACHIE   1812–14   Telford, Hazledine,   *NJ 285 452*
Stuttle, Simpson   by-passed near
Craigellachie off A95

'This bridge of iron, beautifully light, in a situation where the utility of
lightness is instantly perceived.' So wrote the poet Southey in 1819
when the bridge was only five years old. It is now the oldest surviving
iron bridge in Scotland. The modern traveller is given a fine view of its
leap over the river from the modern by-pass. The slenderness of the
centre of the arch is striking, especially if you mentally remove the
railings. Telford was here using a new medium and, a true genius, he did
not make imitation stone bridges. He saw the possibilities of iron—that
it could be cut away and pared down—and he refined his curve to the
sparest possible. To his contemporaries, who had not seen, as we have,
the elegance of reinforced concrete designs, the effect of Craigellachie
was breathtaking.

He placed the bridge on this bend of what the promoters' called a 'rapid and tempestuous river' because the water is here held in check by a cliff of hard Moinian gneiss. In fact, the name Craig Ailichidh means strong rock. Further cut into by the engineers, this cliff is now almost sheer and many stunted trees have grown in the crevices, beeches, pines, elms and rowans. Telford also allowed for floods by placing the bridge on abutments 12 feet above normal water level, and it withstood the phenomenal flood of 1829 although the Spey rose here $15\frac{1}{2}$ feet. The flood arches were swept away however.

The iron was cast at Plas Kynaston in Denbighshire by William Hazledine, an ironmaster Telford regularly used and nicknamed Merlin Hazledine because of his apparently magic skill. The cast ribs were brought here by boat, first on the Llangollen Canal, passing over Telford's amazing cast iron aqueduct at Pontcysyllte in the process, and then by sea and up the Spey. It was erected by William Stuttle, Telford's foreman, as the great man himself was rarely in one place for long. Southey called him the Colossus of roads and said he was 'everywhere making roads, building bridges, forming canals and creating harbours, works of sure, solid, permanent utility'.

The span of the iron work is 150 feet. There are four ribs, each 15 feet apart, and they make an arc of smaller radius than the roadway, which partly accounts for the lightness of the bridge. The spandrels are formed of diamond lattice which also contributes to the delicacy of the design. The castellated rustic ashlar towers—50 feet high and hollow with false arrow slits—which decorate the abutments are perhaps rather heavy, though in spring sunshine they looked fitting enough. The stonework was done by another trusted assistant, the mason John Simpson of Shrewsbury.

The bridge cost £8,200, which was £200 more than the estimate, and the money was found by the Parliamentary Commissioners and local subscribers. It was restored in 1964 by Banff, Moray and Nairn County Councils, and by-passed in 1972. If you stand on the modern bridge and feel it sway as a heavy lorry charges downhill, you can imagine the stresses on Telford's slender arch, even though traffic was lighter. An interesting sidelight on the bridge is provided by a photograph, in the museum at Dufftown, which shows the Spey frozen over and ladies and gentlemen skating below the bridge, watched by several patient, and probably rather chilly, dogs.

Craigellachie Bridge made a strong impression on all who saw it in 1814. One person was the fiddler Will Marshall (1748–1833). He was the Duke of Gordon's factor at Keithmore but found time also to study mechanics and astronomy, to make clocks, to build houses and to play the fiddle. He composed 250 tunes most of them rich and innovative. In

1814, by way of celebration, he wrote a strathspey called Craigellachie Brig. This was played to me by Charlie Menzies of Rosehall and it is a fine, inspiriting tune. At the same time Marshall wrote to his son that the bridge 'is now finished (and is a) beautiful piece of workmanship. I hope you will soon have the pleasure of galloping across it and seeing your sister and friends on the other side of the Spey'—which quotation reminds how vital these bridges were to the communities they served.

At BOAT O' BRIG (*NJ 318 518*) on the B9103, there is both a road and a rail bridge and both replace earlier bridges. A span, over water and marshy meadow, of 250 feet was needed here. Captain Samuel Brown put up a suspension road bridge in 1831 which lasted until 1938 when it was replaced by this single span, steel bow truss. The first railway bridge for the Inverness and Aberdeen Junction Railway was built in 1858 by Joseph Mitchell. His six semicircular approach arches remain. They are built of Elgin sandstone. Mitchell spanned the water with a 70 feet high iron plate-girder bridge which was 230 feet long. It was replaced in 1906 by a single steel truss designed by Head Wrightson. The start angularity of this pair of trusses makes a curious contrast with the harmonious masonry arches and with the soft countryside. Yet another element is introduced by the old toll house, built in the Doric style with four pillars.

OLD SPEY BRIDGE at Fochabers	1806 & 1854	Telford and an unknown engineer	*NJ 340 594* by-passed by the A98 just west of Fochabers

When Cumberland's army were marching towards Culloden Moor and arrived here, they had to wade across. Perhaps the officers used a boat. There was no bridge before the Parliamentary Commissioners started work. One of Telford's problems, in building Highland bridges where none had been before, was to estimate accurately the flood force of the rivers. He took a relaxed attitude to possible failure, saying, in effect, that it was a case of trial and error and his first bridges were not much more than try-outs. It is a tribute to his engineering skill, and to the sensible notice he took of local opinion, that so many of his bridges still stand. This one was a casualty of the 1829 flood.

It spans a 340 feet waterway and was built for £14,880. It was a four span (73, 95, 95, 73 feet), segmentally arched, sandstone bridge using stone from Spynie quarry near Elgin. It had dressed stone voussoirs, a string course, handsome oculi over the side piers and a column on the central one. There is a beautiful watercolour of this bridge, with a rainbow in the sky, in Brodie Castle. Only two spans remain. The fallen

59   Spey Bridge, Fochabers. Another combination bridge

arches were replaced at first by a wooden span designed by Archibald Simpson of Aberdeen, and then, in 1852 (for a cost of £4,000) the three rib cast iron arch you now see was built. This is the remarkable size of 185 feet. There is a square lattice in the spandrels, which has been strengthened with steel cross bracing; and a footpath has been hung off the railing.

The join between the two parts and two styles is uneasy to say the least, and the modern railing does not help matters. Both are fine examples of their type but do not make a satisfactory whole. Down at the waterside the masonry of the earlier bridge can be appreciated and the lovely colour of the sandstone. Notice that the voussoirs are alternately recessed, and that the oculi have courses of dressed stones inside and are ringed by a rope-like circlet, which is altogether finer work than that on the Elgin bridges of a similar style. It is curious how many bridges in this coastal area have oculi. Did the architect of the first set a fashion in Moray?

Walk over the bridge to see the old toll house, now converted and extended. From this point you can appreciate the differing angles and slopes of the two bridges as they cross the river either side of the toll house.

The account of the fall of Telford's arches in 1829 makes exciting reading. Lauder describes the Spey as 'one vast undulating expanse of dark brown water from the foot of the hill of Benagen to the sea, about 10 miles in length and in many places 2 miles broad. The floating wrecks of nature, and of human industry and comfort, were strewed over its surface, which was only varied by the appearance of the tufted tops of trees, or by the roofs of houses to which, in more than one instance, the miserable inhabitants were seen clinging, whilst boats were plying about for their relief.

'By 8 o'clock the flood was 17 feet up the bridge but still its giant limbs magnificently bestrode the roaring stream, which, disparted by the opposing piers, closed round them in perfect vortices, forming a high curved crest from one bank to the other . . . At 20 minutes past 12 o'clock a crack, no wider than the cut of a sword, opened across the roadway, over the second arch from the toll house.' Some men who were crossing had to run for their lives and one had to make a mighty leap over the gap and left one foot 'hanging behind him in vacancy'. Then all the masonry fell 'with the cloud-like appearance of an avalanche'. For a fraction of time the river was 'driven backwards with impetuous recoil, baring its channel to the very bottom, and again rushed onwards, its thundering roar proclaiming its victory, and not a vestige of the fallen fragments was to be seen'.

SPEYMOUTH VIADUCT	1886	Patrick Barnett, Blaikie Bros of Aberdeen	NJ 345 642 walk from Garmouth (on B9015), or from old station on B9104

This huge black metal bridge is an amazing structure to find at the end of a small country road leading to a quiet village and unfrequented beaches. It carried the Great North of Scotland Railway from Elgin to Portgordon, and its size and strength were made necessary by the width and power of the Spey, and by its constantly changing course here at the estuary. You will see that the main span is now over boggy land set with kingcups in spring and cut by many side streams, and the strongly flowing river runs under the right hand access truss; but this could change at any time of violent weather.

The viaduct is 317 yards long overall, and is composed of three sections. The side approach wings are each of three 100 feet trusses, 10 feet high. They are set inside ornate, castle-style stone entries, and supported on circular ashlar piers with rounded granite tops, like plump stone cushions. These piers rest on 14 feet diameter cast iron cylinders filled with stone at a depth of 52 feet below the river bed. The side spans terminate in the same stone embrasures which flank the central section.

60   Speymouth Viaduct with fishermen hauling in a net

61   Speymouth Viaduct, detail of the piers from below

This is a magnificent bow string truss, 350 feet long and rising to 40 feet 8 inches. At the bottom of the arch are two massive open box girders. From these rise vertical spars held by diagonal cross girders of the ladder type. The deck throughout is of metal transverse slats on which the single line rails once ran. Now there is a wooden walkway. Above, the parallel arches are tied with girders at two levels, and, with the rise of the arch, these become farther apart. Crossed girders make the top outer curve of the span. All this detail, which may read as complicated, has inspiring simplicity to the eye. Walking through the viaduct, one is aware of the constantly changing, regular geometrical patterns—diagonals which move against verticals which slide past horizontals, and the whole construction framing the river scenery and, along the coast, the shapely cone of the Bin of Cullen.

In early May there were still touches of snow on the inland hills from which the river comes, but the banks were yellow with gorse and broom. Towards the sea was a sunlit view of the handsome buildings at the mouth. A JCB was levelling pebble banks disturbed by winter storms, and fishermen were casting nets for salmon. One of these men could remember trains crossing the bridge and how Speymouth station, painted yellow, used to win prizes. He told me the Spey was the fastest river in Britain.

It is well worth walking on the riverside path to the estuary (aptly name Tugnet) where there are interesting buildings connected with the salmon fishing industry.

For those interested in railway engineering it would be worth going outside the Highland area to Cullen (*NJ 505 673*) some 10–12 miles on eastwards, where the same Patrick Barnett built huge stone viaducts over the town for the GNSR. They are remarkable structures. The harbour at Cullen and others along this coast were built in conjunction with the Parliamentary Commission roads to facilitate trade and fishing.

CRAIGMIN over the Burn of Buckie	late eighteenth century?	unknown	*NJ 441 621* near Drybridge off A98 in Buckie District

The Highland area has been extended a little to include this extraordinary bridge in this book. Take a right turn near Buckie to Deskry and then another right turn to Drybridge. Go through this small village and straight on up the hill for about half a mile to where you will see Greencraig Farm on the left. Park here and walk along the track in front of the farm.

62   Craigmin Bridge's three tiers of arches

The first thing you see of the bridge is its wavy parapet and the short pillars buttressing it on the inside and its comma-shaped ends. Then you take in the depth and narrowness of the gorge cut by the burn. To see the bridge properly you must clamber down the bank some distance— take care in wet weather—and then the full oddity hits you.

The burn was first spanned by a segmental arch (36 feet 6 inches wide and 34 feet high) of coursed rubble with deep spandrel walls and narrow voussoirs. On this foundation a superstructure has been erected composed firstly of two semicircular arches (28 feet across and 14 feet high) separated by an extremely wide spandrel in which, at a higher level, a fourth arch has been cut. This is segmental and measures 14 feet by 5 feet. Above these three tiers of arches rises the parapet in a curve in which is placed an empty niche. Closer inspection shows doorways into the upper central spandrel where there is a small room lit from above by the topmost arch.

There are few facts to be gleaned about this bridge, but several theories. It almost certainly has a link with Letterfourie House designed by Robert Adam in 1773. The road from the house to the coastal road originally ran this way. Mr Stuart Forbes, who was brought up near here, wrote to me: 'It has been suggested that the lower arch was a bridge in itself and that the upper arches were added on later, a suggestion difficult to prove or disprove because of the skilful blending of the stone work. The reason for the addition was said to have been a mishap to one of the ladies of Letterfourie House which occurred as a carriage in which she was travelling was approaching the bridge. The road had a snake-like bend on the steep incline and as the vehicle attempted to negotiate this bend the carriage lurched and the good lady was hurled into the water below.' It is hardly conceivable that a carriage would have attempted such an incline, but, if the first simple arch had been used only for pack horses, then the theory of an addition later to raise it to a height easier for wheeled traffic makes sense.

Alternatively, it could have been designed like this as a three-tier structure, by an enthusiastic disciple of Adam perhaps, as a thoroughly 'romantic' bridge to be viewed and appreciated from the banks of the stream which might then have been less wooded and threaded with paths in the manner common to eighteenth-century landscape improvers.

It is known that the parapet and niche were added about 1900. The purpose of the cell-like room is also unclear. Mr Forbes suggests it could have been a shelter for Covenanting pickets but this seems unlikely on the property of a Papist landlord. There do, however, seem to be the remains of stone beds. Another idea is that the ladies of the time took refreshments there, à la Marie Antoinette playing at milkmaids, but it would have been an awkward scramble down with baskets, and a rather chilly seat when you arrived.

The road which crosses the bridge used to fork at Greencraig Farm. The right hand road went downhill to the village and to Buckie, coming to the main road at the distillery. Part of its route is still a good track lined with beech trees. The other fork went more or less straight on to the wood you can see. Within living memory this road led to Linn House; the dower house of the estate. Only the first 100 yards now remain. The rest has been ploughed up and the house and stables pulled down. In the wood over the field there is a small stream in a deep earthen gully and this is crossed by a stone bridge which does still exist (*NJ 434 613*). To find it you have to go up the parallel road to the one you are on and look for a smallholding called Back Stripe; this name is on a board tacked to a tree. The bridge can be found by a determined searcher in the wood. (Keep going as straight as possible from the house.) It is a tall, semicircular rubble arch with a splayed parapet, draped with ivy and thick with nettles, still fairly sound, but for how long?

VII

The Upper Dee and Upper Don Valleys

Map 7

If you approach the Dee from Tomintoul, you come first to a section of the River Don which the road crosses at Cock Bridge where little remains of the old bridge. It may be convenient for you to explore the short section of the River Don included in this book first. Take the B973 at its junction with the A939. If you follow this road as far as Deskry you can then go down the A97 to join the River Dee at Dinnet. Apart from the bridges, it makes an interesting drive.

POLDULLIE BRIDGE (*NJ 349 124*), which is just off the B973 at Strathdon, was built in 1715. Whether John Forbes, whose name is on the bridge, was the architect or the paymaster, or both, one has to be grateful to him for providing this beautiful arch which rises with simple serenity over a pool in the young river.

It is set on a natural rock footing, a semicircular arch built of rubble with high spandrels and curving wing walls. The road is approximately 8 feet wide. It lies now on a minor road lined with trees, and stands above two weirs. The water of the pool when I was there was dimpled by fish and flies, and a flock of ewes with their lambs grazed silently in the green field.

To John Forbes of Inverernan we also owe the BRIDGE OF NEWE (*NJ 374 121*) 2 miles to the east because it was his bequest to his great nephew, Sir Charles Forbes of Newe and Edinglassie, that financed it. Newe Bridge was built in 1858 by John Willet, with James Abernethy as contractor. It is a cast iron arch with ashlar abutments and flood arches of two types of granite. The arch has four ribs, each of three sections, and, if you walk underneath, you will see the brick vaults between these ribs and the wrought iron rods tying the ribs together. The cast iron railings are a simple vertical pattern and are 34 inches high. There are no fewer than four ornamental plaques to explain about Sir Charles and his 'grand-uncle'. All the iron work is painted a pale blue which looks surprisingly well, at least it did with the young spring leaves.

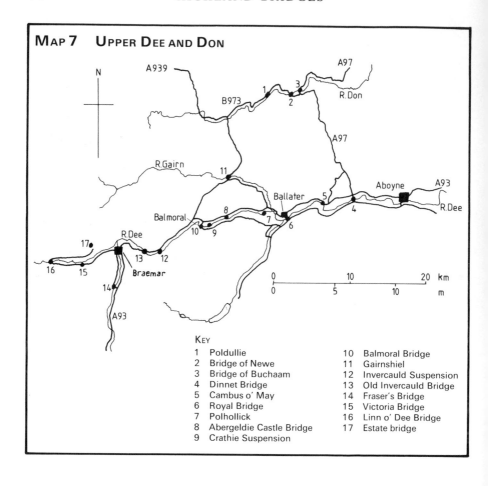

Map 7   Upper Dee and Don

Key

1 Poldullie
2 Bridge of Newe
3 Bridge of Buchaam
4 Dinnet Bridge
5 Cambus o' May
6 Royal Bridge
7 Polhollick
8 Abergeldie Castle Bridge
9 Crathie Suspension

10 Balmoral Bridge
11 Gairnshiel
12 Invercauld Suspension
13 Old Invercauld Bridge
14 Fraser's Bridge
15 Victoria Bridge
16 Linn o' Dee Bridge
17 Estate bridge

63   The perfect circle of Poldullie Bridge on the River Don

Those interested in the Forbes' bridges might like to look at LUIB BRIDGE (*NJ 265 087*) built in 1830 by Sir Charles. It has two unequal stone arches and is altogether of a different period from this iron structure.

The easternmost bridge on this short section of the Don, where it bends round the hills of Cummerton, Corlich and Greenstile, is the BRIDGE OF BUCHAAM (*NJ 391 129*). This has a link with those already seen because its engineer was again John Willet. It was built in about 1858. It crosses the river where the road takes two sharp bends so that it can be well seen as one approaches. At first it appears pleasing with its two elliptical arches and low, graceful lines. However, if you park on the north side and have a closer look, you will see that the parts do not sit easily together. The tongue-like voussoirs with pointed top edges seem at odds with the gentle arch curve, and they are cut off at the top by a rounded string course that is in a different style. The string course in turn upholds a kerb that supports the railings. There is a mixture of stone colours and a mixture of tooling.

Willet's bridge has suffered at the hands of Aberdeen County Council who strengthened it in 1964. The original bridge had a parapet with a coping to match the string course. The spans are 50 feet.

A little less than a mile west there is a second bridge by Willet (*NJ 387 123*) in a different style with three main segmental arches and flood reliefs as well.

64   Cambus o' May

At DINNET BRIDGE (*NO 462 982*) you come to the first of the Dee bridges described in this book. It was built in 1935 by F A MacDonald and Partners Ltd of Glasgow. When new it probably looked handsome in an angular way. It is of reinforced concrete with some structural steel work, and has triangular cantilevers which support a central span with a curved shelf projecting outwards. Discreet decoration has been added. The vertical walls are ridged, the abutments slightly tapered with parallel ridges and there are neat steps down to the river banks with low iron gates. All these details have been carefully thought out and in the engineers' plans, which I have seen, look rather pleasing. Now, however, the concrete is discoloured and stained by ugly, dark mosses growing on it which hide much of the fine detailing. Concrete which has slightly decayed like this is, I believe, one of the main reasons why people will say they dislike modern bridges. I hope that the concrete structures of the 1980s, such as Kylesku, do not stain in this depressing way.

CAMBUS O' MAY (*NO 421 976*) beside the A93 is a white painted suspension footbridge that sits prettily in its leafy surroundings. It was refurbished in 1988 by Kincardine and Deeside District Council having fallen into a partly ruinous state. Built in 1905, it was a gift to the area by a Mr Alexander Gordon who came from Kent. Steel rope cables curve down from the tapered lattice pylons which have ball finials. A lattice truss with sides 46 inches high carries the wooden deck which is

65   Old timber bridge at Ballater, now gone

4 feet wide. The suspender rods are bolted into the top member of the truss and the cable comes down to within 7 inches of this. James Abernethy of Aberdeen designed it.

What completes the charm of this light bridge are the entrances. One side has an unusual swing gate and the other a turnstile with a tall pole. It is clear from older photographs that the roadside entrance has been changed. There was at one time a steep flight of steps up to the road. It is a beautiful and peaceful place despite the road being so close, and there are pleasant walks along the old railway track.

An inscription on the ROYAL BRIDGE at Ballater (*NO 372 956*) mentions two predecessors, one built by Telford. His stone bridge had spans of 34, 55, 60, 55, 34 feet. When destroyed it was replaced by a wooden one designed by Joseph Mitchell and probably based on a Telford design. The present stone bridge was financed by the County Road Trustees and opened by Queen Victoria in 1885. Jenkins and Marr were the engineers.

It is a long bridge and the parapet comes down in long steps which mask the rise of the roadway. The abutments are finished with round topped pillars. The four segmental arches are on narrow piers with rounded cutwaters and the central pier is extended up to a square refuge with some ornamentation. This central elongation and squaring off gives a curiously elegant air to what is a solid enough bridge.

It is close to the town square, but, apart from the hotel, the banks are tree covered and no buildings are evident. You can walk along a path and have a look at the cutwaters (one is on dry land normally) and also see the coursing on the inner faces of the piers, like closed arches. Beyond are the ruins of some handsome stables.

Two miles west of Ballater is another suspension footbridge by James Abernethy and Company, also the gift of Mr Gordon. POLHOLLICK (*NO 344 965*) is in a more open situation and perhaps for that reason was built a little more sturdily and with less charm. The lattice towers are finished with a tooth-edged cross girder above a metal arch. The cables, of steel rope, are stayed by cross bars that give a ladder-like effect from a distance. The abutments are lozenge shaped. It was built in 1892.

The next bridge, also a suspension footbridge, was built in 1885 and is in a much poorer state of repair. ABERGELDIE CASTLE BRIDGE (*NO 288 953*) is 1½ miles east of Balmoral. It was built by Blaikie Brothers at Queen Victoria's expense for the benefit of her guests using Abergeldie Castle. It is private and you cannot walk on it; but its rusty state suggests no-one uses it. Two wirerope cables and iron rod suspenders hold the wooden deck. The lattice towers are linked with a pretty iron work bar. The railing is a light woven wire and steep steps lead up to the gate. Beyond is the terracotta coloured castle with its bell turret and gold vane. In May the banks were white with wild cherry blossom under the dark pines.

A river valley often has a type of bridge that is typical of it. In the case of the Upper Dee it is the suspension bridge that reigns. The earliest—built in 1834 and renewed in 1885 at the Queen's expense—is CRATHIE SUSPENSION BRIDGE (*NO 266 942*) on the B976 half a mile to the east of the castle. Built by Justice Junior and Company of Dundee, it was then the main way across the river to the castle, and so was made wide enough to take carriages. This extra width gives it a special charm, particularly if you imagine a horsedrawn carriage with ladies wearing crinolines and carrying sunshades.

The abutments are solid with rounded cutwaters and a flood arch on both sides under the shorebays. The chain is not, in this early and unorthodox example of a suspension bridge, a metal rope but a succession of paired flat links. Four diagonal rods ray out from each tapering wrought iron pylon. As you walk over the bridge, you can see these rods vibrating gently; and, from the bank, you see that they continue below the deck to provide bracing. Below are truss rods providing extra support. It is an unusual bridge by a man who was experimenting with suspension methods. Another bridge he built is at Glenisla in Angus. This one was renewed by Blaikie in 1885, but what his work entailed is not precisely known.

The white paint increases the feeling of lightness, and the pattern of

66   Crathie Bridge, showing the swing gateway

decreasing circles over the flat arches, along with the swing entrance
gates, adds to its picturesque looks. From here you have a view of the
red-tiled steeple of Crathie church, looking remarkably English—
probably because Queen Victoria had a hand in its design.

BALMORAL BRIDGE	1854–57	I K Brunel	*NO 263 949*
		R Brotherhood of	at the Castle
		Chippenham	entrance

After the romantic suspension bridges, Balmoral Bridge comes as a
surprise; and even more so when you learn that the great engineer of the
Tamar Bridge designed it. The reason was royal patronage. Prince
Albert felt the need of a new approach to the castle and he commissioned Brunel.

67   Balmoral Bridge, built by I K Brunel

For some years the fact that Brunel had designed this bridge was forgotten. However two researchers, Angus Buchanan and Stephen Jones, have established that he was its engineer. He was first approached in 1854 and submitted several designs. When this one was accepted, he recommended a firm from Wiltshire to cast the iron work. (I wonder what Scottish comments were passed on this.) As Brunel was too busy a man to oversee its erection, the local factor, Andrew Robertson, who was also the doctor, took on this responsibility and saw to the building of the abutments and the provision of deck planking. However there were delays and the bridge was not in use until 1857—somewhat to the royal displeasure.

Nor was this all. It appears from a letter in which Brunel defends his design that the Queen certainly, and possibly the Prince also thought it too plain. 'Not extremely ornamental' was how the criticism was phrased and Brunel commented, 'I fear your expression . . . implies something very much the reverse.' He says he aimed at a design 'of perfect simplicity'. thinking that making sure there was no unsightly or offensive ornament 'is always a great first step'. No doubt the Queen, whose name has become a term to describe over-elaborate ornamentation, did not appreciate Brunel's elegant functionalism. Certainly the bridge is passed over in her diary in ominous silence whereas she wrote enthusiastically about the opening of the bridge at the Linn o' Dee. It was probably client dissatisfaction which led Brunel's son to omit reference to this bridge in his life of his father. In addition, the contractor,

68   Detail of riveted lattice side of truss, Balmoral Bridge

Brotherhood, attempted later to pass it off as his own; and Brunel never asked for a fee. All these factors may have led to his authorship being overlooked.

The bridge is a wrought iron plate girder construction, probably the earliest in Scotland. The span of the two riveted girders is 129 feet, and the road (tarmac and pine planking) is 13 feet wide. The girders are pierced with a simple diamond pattern in which rivets and plates are used effectively as part of the design. The granite abutments are set on rocks. Their footing caused problems and led to some of the delay. They are rustic ashlar for the most part with smooth dressed rectangular tops and parapet. The splayed approach wall has a most satisfying curved top.

The only other bridge Brunel designed at all like this one, it has been suggested, was for the railway in Bengal—a far cry from the chill, green hillsides of Deeside.

I admire the strength and bold simplicity of this design as a whole and in its detailing. It is not altogether in harmony, perhaps, with the man-made, rather gothic, surroundings; but it suits well the natural scenery of rocks, trees and water. It has a spare functionalism and solid plainness that belongs more to our century than the last. It is now painted a greenish turquoise—not a happy choice. The colour is too like, and yet quite unlike, the colour of leaves or of water. Brunel himself wanted it painted 'some simple sober but warm brown tint', and with the underside and the outside contrasting shades. Maybe this could be done when next it is painted? I noticed that the visitors streaming over the bridge to the castle never gave it a glance.

From here take the minor B976 which crosses the high moors to Gairnshiel, a fine short drive along the line of the eighteenth-century military road.

GAIRNSHIEL BRIDGE (*NJ 295 008*) is a steeply humped and handsome stone arch over the River Gairn and is best seen from a short distance away, as then its grace and dignity can be fully appreciated.

It was built in 1751 by Major Edward Caulfeild as part of the military road running from Blairgowrie up Glen Shee to Braemar, then via there and the Lecht to Grantown and eventually to Fort George on the coast of Moray—a formidable undertaking. It was built in part by five companies of the 33rd Regiment of Lord Charles Hay's, as a stone by the well of Lecht says (*NJ 234 151*). The bridges, however, were not built by soldiers but by civilian contractors.

Gairnshiel arch is 56 feet 8 inches in span, built of coursed rubble with thin voussoirs and a parapet edging of thin flags. The parapet is 41 inches at the top but slopes down to as little as 8 inches. There is no ornament and none is needed. The simplicity of the arch perfectly complements the grand lines of the countryside. Small larch trees grow on steps of rock at the water's edge. In May there were Highland cattle grazing in one field, ewes and their lambs in the next.

During the half hour I spent here five coaches came over. Because of the severe hump the passengers had to leave the vehicle and then the driver took it gently up until it straddled the top and tipped over the far side. I asked myself what sort of strain this frequent manoeuvre was putting on this almost 250 year old structure.

Back on the A93 Deeside road and driving towards Braemar, the next bridge you come to is INVERCAULD SUSPENSION (*NO 197 908*) built in 1924 by James Abernethy and Company. This is the last of the decorative footbridges over this stretch of the Dee. It also is private and similar to Polhollick. Only about one mile west of here you pass Old Invercauld Bridge which is partly veiled by trees. Do not miss it.

OLD INVERCAULD BRIDGE	1752	Caulfeild	*NO 186 909*
			beside the A93 12 miles
			east of Ballater

The tollhouse (of later date) on a sharp bend may help you to pinpoint this bridge. It is possible to park under the trees nearby, but not easy. Paths through the wood lead to the river. Here I met a man who was watching his wife fish, his own cap torn and unravelled by the many flies

69   Old Invercauld Bridge

he had hooked into it. He thought this bridge the finest in the Highlands.

It is a six span, hump-backed bridge of rubble, each span being a different size. Looking it from the downstream side, the largest is the third from the right bank, under the highest point of the roadway, but the second is a fair span too. The fifth and sixth, which are hidden by trees and on completely dry land normally, are very small, the last measuring only 10 feet across and 54 inches high. These two are effectively flood arches, whereas the first on the far bank is not. This asymmetry in no way detracts from the bridge's appearance, rather it enhances it, as do the unusual tall triangular cutwaters which have roofs of large flags. It stands on a rocky bend in the river with reefs of rock across the flow. Caulfeild had to do some blasting to prepare the site, and it is said a man was killed during this operation.

It is clear that the bridge has been repaired at various times. The stonework of the south side wing walls and some of the cutwaters is quite different from the older parts, and the voussoirs have also been changed, but the two main arches (that is numbers two and three from the right, the south, bank) are probably more or less as Caulfeild built them. The parapet is high at the hump, a good 50 inches but it comes down to 23 inches before the ends which are splayed on the north side. It is 17 inches wide and topped with flat stones of different sizes. From the top of the bridge you have a wonderful view of the river and the mountains. In May the eastern cliffs were still snow-edged, and there was a delicious scent of resin from the pines. You can also see from here, turning upstream, the bridge commissioned by Prince Albert in 1859 when this old bridge was taken out of service. New Invercauld Bridge has three main spans with oculi in the spandrels. It is not unhandsome but suffers from its proximity to the bridge built one hundred years before it. It has proved difficult to find the name of its engineer but William Smith, the architect of Balmoral, may well have been retained for the bridge also.

Old Invercauld is a national monument in the care of the Government. It majestically spans the rocks and the river, its pale stone reflected in the river's ochreous green water. It looks so at home in its surroundings that it might be a product of nature rather than man.

FRASER'S BRIDGE over the Clunie Water	*c.* 1752	Caulfeild and others	*NO 148 864* to the right of the A93 going south beyond Braemar

This bridge is also part of the military road built in the early 1750s but a very different thing from Old Invercauld. It is on one of the more dramatic sections of the road leading to the Devil's Elbow and the Spittal of Glenshee. Starting in 1748 various troops were used—General Blakeney's Regiment, General Guise's Regiment and one hundred men from Lord Viscount Bury's Regiment.

The bridge you see has been much repaired, although some of the arches and pier work may be original. In 1832 a road engineer said it was 'somewhat crazy, its parapets swayed from the perpendicular and its mortar not in a good state of adhesion'. It was repaired then and later when Prince Albert and Farquharson of Invercauld remade this road.

The bridge is 100 feet long and has two unequal segmental arches of uncoursed rubble, the east one being 30 feet. It springs at 2 feet, low to the water, and rises 7 feet 9 inches above the springing point. The west span is no longer over the river so that the cutwater has become a buttress. The voussoirs are narrow and uneven, some extending up into the spandrels. On the flat coping stones are some masons' marks and initials, probably the Victorian men. The first builder, employed by Caulfeild, was James Robertson of Dunkeld.

Fraser's Bridge, or the Bridge over Clunie Water as it is sometimes called, spans the river between grassy banks in a wide, bare valley. If you drive over it you can return to Braemar on the minor road past the golf course and church. This is the line of the military road. On this stretch of the river are several small wooden bridges, one of which is particularly goodlooking with vertical rails.

70   Fraser's Bridge over Clunie Water on the military road south of Braemar

From Braemar you can explore the upper part of the Dee. Approximately 3 miles from the town is VICTORIA BRIDGE (*NO 102 896*), a highly ornamental three span iron girder bridge and a surprise in the quiet landscape of this wide valley where the river loops through the gravel beds. The bridge is an iron truss with 62 inch high ornamental sides and arched entrances, one inscribed Victoria 1848 and the other Edward VII 1905. The first bridge here was built in Victoria's reign and the present one in Edward VII's; the railing is from the first bridge. It is this railing which is chiefly memorable, especially when the sun throws its shadows on the roadway. There are dressed granite abutments and two oval piers with wing walls curving down and round and finished with square ends. Altogether, it is a bridge of high and finished workmanship. To complete the picture there is a pillared lodge at the south end built to resemble a toll house. Old larches and spruce and pine cluster on the banks.

71   Victoria Bridge. Note the 1848 railing over the 1905 truss side

LINN O' DEE BRIDGE   1857   A S W Reid   *NO 062 897*
                                                        at the head of the road

If Victoria Bridge surprises, this one is even less to be expected at a spot where the road turns back from the wilder tracks across the mountains. It is a highly finished, elaborately detailed stone bridge of pink and grey granite with a very unusual pointed arch; and set down on the rocky sides of the river where it falls dramatically over a series of rock steps in a narrow gully. Take care when approaching the edge. It was here that Byron, as a young man sent to recover from an illness, slipped and nearly fell to his death.

There is a denticulated string course, moulded voussoirs, small ornamental pointed buttresses and a castellated parapet with a coping, rather than a flat top, which extends on to deep abutments. An ornamental plaque tells that it was built at the cost of the fifth Earl of Fife. Queen Victoria attended its opening and spoke warmly of it.

It has a vaguely ecclesiastical air, and would be more at home in a cathedral close than in the wilds. Yet its very unsuitability for its surroundings, which seem to require something much simpler and plainer like Fraser's bridge for instance, adds to the attraction of place.

If you continue on the road round to where it ends at a rather rackety wood-decked bridge at the Linn of Quoich, you can take a short walk up to the **PUNCH BOWL ESTATE BRIDGE** which is in beautiful surroundings

72   Linn O' Dee, opened by Queen Victoria

and where you may see red squirrels in the trees. The bridge (*NO 115 913*) is no longer safe and looks as if it has replaced a former stone bridge, judging by the abutments. It is not remarkable but is typical of the sort estate factors had built. It spans the impetuous stream cutting its way between stone walls where smooth round holes have been gouged over centuries in the rocks. The bridge is wooden, a beam put across from the old abutments and two diagonal stays making a triangular support below. Beside the bridge is a stone building its windows blinded with boards. From here you can walk up into the mountains, but must be sure to have the right equipment.

On the road back to Braemar from here there are two other old stone arch bridges (at *NO 070 899* and *NO 086 892*) and both cross streams that are the starting points for mountain tracks up Glen Lui and Glen Ey. It is an exceptionally beautiful area to explore, and full of wild deer which browse unconcernedly in the fields. When I was there a man practising his golf shots hardly bothered them.

VIII

The A9—Killiecrankie to Newtonmore

Map 8

KILLIECRANKIE VIADUCT   1863   Joseph Mitchell   *NN 917 625*
beside the Gorge   footpath downhill
from Visitor Centre

Joseph Mitchell was the son of Telford's deputy in the Highlands, one John Mitchell, a mason from Forres. He was known as Telford's Tartar, so hard and fiercely did he drive himself and others. His son took over his post and apparently his work ethic too, judging by the number of undertakings he had a hand in. Towards the end of his forty-five year long career, he started building railways when the mania for them took hold of the country. (At one time in Scotland there were 146 different railway companies.) The Inverness and Perth Junction Railway was the first Mitchell built and a formidable challenge it was, not just the engineering, but also in raising the necessary funds, and pacifying the landowners through whose properties the snorting monsters were to run. Mitchell's *Reminiscences* make interesting and amusing reading.

This viaduct is 510 feet long and 54 feet high, and it cost £5,730. It is handsomely curved—the curve is that of a circle with 440 yard radius—following the bend of the river and built of yellowy-brown and grey stone with brick vaulting. There are ten spans of 35 feet. To harmonise with the rugged scenery it has been given castellated towers at both ends and in the centre, and a denticulate string course. Beyond the viaduct arches the masonry has been continued with a turreted retaining wall to a deep barrel-vaulted underbridge crossing a burn. This fine little arch makes a sharp contrast in point of technique, care of execution and looks with the road underbridge you have passed coming down from the carpark. They both perform the same task of bridging a small, deep fissure, but Mitchell's arch and masonry are things to look at with pleasure.

MAP 8 A9—KILLIECRANKIE TO NEWTONMORE

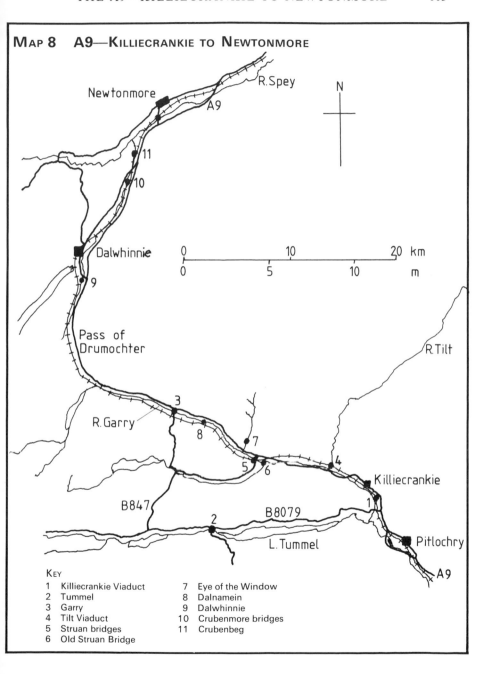

KEY

1	Killiecrankie Viaduct	7	Eye of the Window
2	Tummel	8	Dalnamein
3	Garry	9	Dalwhinnie
4	Tilt Viaduct	10	Crubenmore bridges
5	Struan bridges	11	Crubenbeg
6	Old Struan Bridge		

73   Tummel Bridge, photographed early this century. Mendelssohn stayed here in
1829. Courtesy RCAHMS

The section of the A9 which now by-passes the National Trust carpark was the final stretch to be built in 1986. In effect it is a bridge, 656 yards long with forty-one spans each 17.4 yards high, which strides across one of the most difficult sites on the whole road north of Perth. I believe Mitchell would have appreciated it. In the small museum at Blair Atholl, called the Atholl Country Collection, there is in the entrance porch a fine wide photograph of the work in progress. It shows the piers, the earth cuttings and the embankments.

From Killiecrankie take the beautiful B8019 towards Rannoch, past the Queen's View to Tummel Bridge.

TUMMEL or	1730 General Wade and	*NN 762 592*
CANAGAN BRIDGE	John Stewart	junction of B8019
		and B846

This fine bridge was built for Wade by John Stewart of Canagan (hence its alternative name) for £200, plus a repairing duty for twenty years. Hardly a contract that would be acceptable over such a stormy river today. Stewart built well and strongly, for his bridge stands 260 years on. The contract does not specify the design or even the exact location. It is to be a double arch over the River Tumble (*sic*) with a span of 42 feet 'between landstools' (that is abutments), 12 feet wide and with a 3 feet parapet 'coped with good flag stones'. It is to have sufficiently long approaches as 'to render it easily passable for wheel carriage and cannon'. For this Wade paid out of his Government funds £50 down and £150 on completion. There are few of Wade's contracts still existing and one wonders whether all 'Wade bridges' are almost entirely the work of other men. If this is the case, and it probably is, the innumerable stone arches pointed out to me as 'the Wade bridge' or 'the General's Bridge'—as if there were only one General and only one bridge—are even less aptly attributed.

That it is a fine bridge may not be immediately apparent to you, for a green metal truss bridge has been placed far too close to it and you cannot look at the arch without its being bisected by this girder. The twentieth-century tide has washed up against it in the form not only of unsympathetic concrete restoration but also houses, chalets and a power station. Unclipped brushwood is thick on the banks where there is litter underfoot. Traffic also stops one from looking in peace at the bridge. It is, however, worth the mental effort required to imagine it as it was, alone in a rock strewn hillside.

It has a main arch of 55 feet set high with a humped roadway. The parapet is 4 feet in the centre and reduces to 2 feet 6 inches. The road is 11 feet wide. There is a segmental flood arch on the north side, unusual in being set at an angle of 20 degrees to the main arch. Unfortunately an ugly buttressing hump has been put on the upstream side, presumably as a baffle to flood water swirling round the new bridge, and this further inhibits appreciation of the line of Canagan's bridge with its curving wings. There is a plaque saying that it was built in 1733 and renovated in 1973. I think the date of 1733 is inaccurate. In any case, the notice is sloppily executed and seems typical of this place which has allowed a beautiful and historic bridge to be marred.

The nearby farmhouse which appears in old photographs is probably the original Kinghouse on the military road. Mendelssohn stayed here in August 1829 and he wrote:

> The storm howls, rushes and whistles outside, slams the door and opens the shutters . . . The dining room is large and bare. On one wall the wet trickles down. The floor is thin and below us in the servants' quarters there is noisy talking, with convivial singing, laughter and the barking of dogs. There are two beds for us, with purple curtains; on our feet, instead of English slippers, Scottish wooden shoes. There is tea with honey and potato cakes. On the narrow winding staircase the maid met us with schnaps. Only gloomy driving clouds in the sky. And yet, in spite of the noise of wind and water, in spite of the servants' clamour and the rattling doors, there is a stillness, very lonely. It is a Highland inn.

From here go on to Trinafour taking the long straight road north on

74   Tilt Viaduct with baronial portal

the exact line of Wade's road with fine views of the whaleback of Schiehallion, snow flecked when I saw it in November. At Trinafour is another Wade bridge, rather ruinous and difficult to see. The moorland road continues north to Dalnacardoch, and, for the most part, follows the military line. Above Trinafour there are traverses and then Errochty Dam and the reedy Maud Loch. The red disks on posts here are to mark the road in snow. At the top are fine views and a sign saying 'Spare the Sheep'.

These military roads were built across unknown hills. The year before work began a survey map would be made, in the early days by amateurs. The first step in actual construction was to dig out the peat and loose material and pile it in banks on either side. These can often still be made out. Where the base was flat, firm rock, little else was done, as has been discovered by recent excavations near the line of the modern

A9. In other places the rocks needed for road surfacing were usually lying around in abundance, and had only to be split and pounded down into a bed. The rate of building expected was about $1\frac{1}{2}$ yards per man per day. Split by gunpowder and sledge hammer, the rocks were laid in progressively finer layers and topped with gravel to a depth of 2 feet. This was the ideal, perhaps not always attained. The gravel needed frequent topping up due to water erosion, although, wherever possible, back drains and paved cross drains were put in. (Tar is not mentioned as a road surface in the Highlands until 1857.)

This particular road crosses the River Garry at Dalnacardoch by the GARRY BRIDGE (*NN 727 701*) built in 1730 by Wade. It has been coated with harling as a preservative measure and this somewhat detracts from its appearance. In old photographs the attractive rough texture of the coursed rubble is seen. However, the simplicity of the single arch framing the waterside can be admired. Hydro-electric schemes in this area have much reduced the flow of the Garry so that the tumbled grey rocks are more exposed than they would have been in the 1700s. the span is 47 feet, the abutments are built on rock, the voussoirs narrow. One wing wall is longer than the other, with a slight splay at the end. It is approximately 101 feet long, and it is typical of military bridges in being without decoration. When I visited it in November there was a thin, clear ice sheet encroaching on the water from the rocks, and a hawk was hovering. An Intercity train passed and the hum from the A9 could be heard. I held in my mind the modern transport and the thought of the soldiers in their cold discomfort building roads by hand, and being paid one shilling a day.

TILT VIADUCT	1863	Joseph Mitchell	*NN 873 652*
over the River Tilt		Fairburn & Sons	park in Blair Atholl
		of Manchester	and walk down the
			riverside

Mitchell, bringing the first railway track north in the 1860s for the Inverness and Perth Junction Railway, bridged this wide river entry into the Garry with a positive statement of strength and durability. The castellated portals, as if the engine steamed through a portcullis, seem to our eyes rather over the top and even Mitchell, who liked the baronial style, said it was made 'somewhat more ornate than was otherwise necessary' because of problems with the Duke of Atholl. Poor Mitchell had difficulties with His Grace who, when the line was first proposed, declared bluntly he objected to all railways in the Highlands and would only agree to this one if it was not detrimental to his estate. Mitchell staked out the line and marked it with white flags, then accompanied the Duke in his carriage for a viewing. The Duchess was also present and seems to have smoothed matters out. Mitchell comments that it was no wonder this charming woman was a favourite of the Queen's.

The bridge itself is a single, diamond lattice truss span of 150 feet supported on stone abutments. It has been given additional steel bracing since, in the form of eight triangular arms placed on each side. It is interesting here to walk under the bridge—the river is usually low— and look up at the elaborate meccano work of its riveted and bolted construction. Mitchell tells us that the girder is wrought iron, and the abutments laid on a platform of 6 inch timbers 3 feet down in the river bed and secured to piles, its height above the river 40 feet and its cost £25. 7s. 9d. per linear foot. The total length is 256 feet.

Tilt Bridge is a name older than this viaduct. Behind the town there is a road bridge of Tilt (*NN 876 664*) at a convenient narrow point that has been a bridging place for centuries, although at one time it was known as Black Bridge. Wade built a stone bridge here to replace a wooden one, and his was superseded by another that has recently suffered modernisation. Just above it is a pedestrian archway crossing the road from one part of the Castle grounds to another.

At Struan there are three bridges over the Garry (*NN 802 657*), a road bridge arched over by two viaducts. The road was built first, probably in 1765, but has clearly been altered. In 1863 Mitchell took a single rail over road and river on a castellated viaduct. He also had difficulties over this with the Duke of Atholl who did not want his waterfall and plantations spoiled. He says he overcame the problem by 'spanning road and river obliquely at the narrow point by a three arch bridge'. His bridge has one segmental and two semicircular arches, and is built of stone with brick vaulting, now patched with stone in places. Next to it, and so close that the turreted decoration on Mitchell's bridge had to be cut into, is a second viaduct put up in 1899 in order to make this section of line double track. This is a four span bridge with two central iron truss spans flanked by masonry approaches on massive stone piers and abutments. The main truss has overhead stabilisers, three rounded and decorative, two squared and doubtless later. The iron work was cast by Alexander Findlay and Company of the Parkneuk Works in Motherwell. The two bridges are cunningly linked and, on the right bank, there is a barrel vaulted passage for pedestrians. The rocks are strikingly pale under the clear green water which rushes under the bridges. It is a place that obviously needed a bridge for safe crossing, and, in the Clan Donald Museum at Bruar Falls, there is a copy of a map made in 1756 by John Leslie which notes 'a bridge necessary' here (and elsewhere).

If you continue down the B847 and take the first turning left to Old Struan you will come to OLD STRUAN BRIDGE (*NN 809 653*), a pleasant stone hump-backed arch over a tiny gorge through which the Errochty Water flows. Above the bridge is the parish church in the heart of Robertson clan country. The bridge was probably built in 1765 and replaced a succession of wooden ones constantly destroyed by floods and constantly repaired to allow access to the kirk. The elders raised £50 sterling, half by subscription and half donated by local landowners. It was probably built by a Dunkeld mason and is simple and undecorated with the parapet coping stones now riveted together.

EYE OF THE WINDOW	1728 Wade	*NN 791 668*
over the Allt a'		reached on foot from
Chrombaidh		Bruar or Calvine

Wade built the earliest recorded road through the Drumochter Pass—though it is not impossible that the Romans were here before him—and for the most part his line is buried under later roads, due to the narrowness of the passage; but this charming, restored bridge is on a section which lies above the modern roads and is partly preserved as a moorland track. You can walk there and back from Bruar Falls comfortably in two hours. It is 2 miles there, uphill mostly.

The start of the path is not marked either by a finger post or on the map of the Falls which could perhaps be rectified. To find the path, go up the falls path under the railway bridge. You will see various grassy paths going off left. The best marked and widest is a good 100 yards up from the railway bridge, and it goes straight, crossing a burn and then rising to join a stoncy track up through the wood. You emerge, pass a gate and strike off across the moor. At the point where you come level to the wood above Calvine do not turn right uphill with the stoney track, but keep straight on a level grassy track beside the field wall. This takes you eventually downhill into a second wood and you find the bridge almost at once. It has a grassy roadway and is overhung by trees. The steeply falling Allt a'Chrombaidh cascades beneath.

Its charming name is a translation of the Gaelic Drochaid na h-Uinnelge. It is a semicircular arch of roughly coursed rubble with thin, irregular voussoirs. One abutment is set firmly on the rock shelves of the river bank but the other hardly exists. Wade has set the arch down on the rock itself. It is a beautiful place where a walker is inclined to rest rather than press on. The wood is of pine and larches which, in October, were golden and set off by the scarlet bouquets of rowan leaves and berries. The peaty water, creamy with foam, careered down the rock falls leaving foam puffs slowly turning in the slack water bays.

75   Dalnamein Bridge on the old A9 south of Drumochter

The military road continues for some distance after this bridge and curves down to meet the old A9 between Clunes and Dalnamein. On this stretch is another Wade bridge over the Allt nan Cuinneag (*NN 781 676*). At Dalnamein it crosses a stone Parliamentary bridge which was by-passed in the 1920s building programme (*NN 754 697*). Beside it is an indecipherable milestone.

DALNAMEIN BRIDGE over the River Anndeir	1928 Owen Williams and Maxwell Aryton	*NN 756 695* on the old A9

At Calvine take a turning marked as a cul-de-sac between the A9 and the turn to Struan. This road was in use as a trunk road until the mid-1970s and runs beside the river with trees overhanging now and occasional deer crossing. At Dalnamein the concrete bridge has had half its roadway taken up but is safe to cross.

I find this one of the most pleasing of the bridges Owen Williams built in the 1920s. It is an unusual design with two arch ribs and square columns rising from the arches to support the deck. At the spring of the arch the columns are in pairs. There is a simple straight parapet decorated only with vertical slits; the abutments are large and stepped back.

The bridge is skewed which has the effect of putting the arch ribs and the columns 12 feet out of alignment. The cross beams are therefore diagonal. The columns have unconventional capitals which project as

76   Dalnamein, detail of column brackets

bracketlike arms. Three of these brackets on each column head are short and they support the roadway which is cantilevered out 5 feet on each side. The fourth arm on each head, the inner one, is extended into a beam supporting the deck and linking with the capital head of the opposite column, which is 12 feet along the bridge. This can be clearly seen if you go down the bank and look up, but is hidden from above by the jutting edge of the deck. It is well worth going down as you can get right under the bridge and see it in detail. Engineers are interested in 'the complex structural support for the slab combining beam and flat-slab construction', to quote from a book on William's work. His buildings are coming to public interest after a period of neglect and it is hoped that this bridge will be repaired and so retained, particularly as the old road makes a pleasant quiet escape route from the busy A9.

The slender lines of the arches and the unaligned columns, through which the light falls, give the bridge a pleasing lightness despite the massive simplicity of the concrete. The double columns at the springing are the tallest. From that point they shorten upwards in both directions which adds to the poise of the structure. Apart from the parapet there is little in the way of solid wall—a strong contrast to the heavy bridge over the Findhorn by the same two men. In a photograph taken when it was new, the concrete is pale and the detailing crisper and much more telling.

Immediately on the far side of the Drumochter Pass you come to an old restored military road bridge.

DALWHINNIE BRIDGE   1728 approx   Wade   *NN 639 828*
over the River Truim   just off A9 at
                                                         Dalwhinnie

The bridge no longer has a parapet but can safely be crossed. The main span is 30 feet, and the small flood arch 10 feet and only 5 feet high. The river here flows between less rocky banks and altogether more temperately than is usual with Highland rivers so that the span is flatter than usual. It is particularly interesting in that it shows more clearly than many how the early bridge masons constructed their arches. You will notice that the voussoirs do not radiate round the arch ring in a strictly precise way. More of the lower ones are nearly horizontal and more of the upper ones vertical than the best practice would permit. The authorities seem to agree that probably these rings were built without mortar at first, the lower stones corbelled out at the bottom and the upper ones perhaps jammed together by flakes of stone and then the mortar poured in from above. This configuration of the voussoirs indicates in the Highlands that the bridge is early, that it probably pre-dates the Telford roads, although, of course, in some areas old methods may have continued into the nineteenth century. These early bridges have almost invariably segmental arches, some quite shallow, and it is clear that such a shape would be helpful in building by the method of assembling stones on the centering without mortar. I can refer those interested in these technicalities to works by Ted Ruddock and G R Curtis, to both of whom I am indebted.

Near this bridge is a hotel which started life as a Kingshouse on the military road which here ran farther west than its modern counterpart. If you take the right hand turn on to a minor road running parallel to the river you will be led to the next bridge described, and also see several others, of different types, crossing from one pebbly bank to the other.

The CRUBENMORE BRIDGES (*NN 676 913* and *674 915*) should really be considered along with Crubenbeg Bridge less than a mile farther north because they were each built after a realigning of this arterial road. Wade took his road over the Truim at Crubenbeg in 1730 but later in the century it was substantially rerouted at different dates. At Dalwhinnie it was subject to floods and here it tended to be 'blown up with snow' in winter. The Government and the army both provided funds for this work, and different sections were done, and redone, at different times so that it is difficult to be certain when this bridge was built. It is connected with the improved road the Parliamentary Commissioners built after 1802 but I do not think it can definitely be claimed as a Telford bridge. It may well have had an army engineer.

It stands beside a cottage with a white gate which gives access to the

77   Owen William's small bridge at Crubenmore on the old A9

old road leading to the bridge, which was restored in 1975 and won a prize. The very fact that it has two wide spans of 33 feet each and a solid pier set in water would suggest it is of later date than the other military bridges in this area. The segmental arches have triangular cutwaters and are built of very unevenly sized stones, tufted now with grass. The schist flags making the voussoirs are well seen in the vaults which are easy to walk under. The abutments are set on the sloping grass banks of the river and extend on the east into a long curving wing wall. If you cross the bridge, where in July ladies bedstraw was growing in the grass, you will find an old iron swing gate giving access to a riverside path, and have a close view of the fine drystone wall. It is a bridge with no finish at all, a sturdy but rather crude construction, yet handsome for all that and suited to the wide, high valley with the scree tops and crags above.

Just to the north the modern road crosses the river on the second Crubenmore bridge, built in 1926 by Owen Williams, with a span of 72 feet. It has curious faceted arch rings which extend into the piers and abutments and form pointed cutwaters, somewhat like a boat's keel or the snub bow of a submarine. The semi-hexagonal bay in the centre of the parapet over the pier could be the sub's conning tower. Behind this angled shuttering are reinforced concrete arches. The bridge is on a bend in the river and so has to be skewed. There are similar small bridges near Dalnacardoch over the Edendon Water (*NN 716 706*) and at Loch Alvie (*NH 871 093*), also by Williams. Both the Crubenmore bridges have now been by-passed in the latest main road north which avoids crossing the Truim.

78   Crubenbeg, Wade's bridge over the River Truim. Note springing off rock

CRUBENBEG BRIDGE (*NN 680 923*) is to be found just before this minor road (the old A9) joins the new road. Coming from the other direction the turning is signed Dalwhinnie Tourist Road. The side road down to the bridge is steep, narrow and short so it may well be better to park at the top where there is a layby and walk down—three minutes at most.

General Wade built this bridge in 1730 and so it was the earliest of the many crossings of this river Truim. It is typical of its time in being seated on the rocks and single arched so that no difficult pier work under water was needed. The main span is a semicircular arch of 30 feet and on the right bank there is a flood arch 10 feet wide and only 5 feet high—the same dimensions as the flood arch at Dalwhinnie—and there is a triangular cutwater, or buttress rather, on the upstream face. The parapet is low, only 17 inches at the highest point, going down to 13 inches and has ends curving through a quadrant.

It is built of rubble, the stone being schistose. This has many mica flakes in it and splits thinly, though not smoothly, thus making pieces which are flattish but do not have the clean surfaces of sandstone and slate. You will see the surfaces are fairly rough and of all shapes and sizes.

The bridge and its setting is very picturesque with little step waterfalls between the irregular lichened rocks, and little trees which sprout there. Close to the bridge is a bent old larch and downstream a group of pines below a field full of old roots and stumps. In a hot July the short grass was burned down and had purple thyme flowers in it. Near the water

Queen Anne's lace was growing and the silky water was flecked with foam spots which cast tiny moving shadows on the rocks below. This would be a restful place to break a journey.

The magnificent modern bridge at Newtonmore is described in the chapter on the Upper Spey, see page 62. It is about 5 miles north of Crubenbeg.

IX

Roads south, west and north of Beauly

Map 9

The pleasant small town of Beauly at the head of the Firth stands at the intersection of various routes. There is the road south and west down Strathglass and into Glen Affric, the road through the western mountains to the coast, and the firthside road north towards Wick. This chapter follows these three routes.

1 *The Strathglass road A831*

LOVAT BRIDGE   1811–14   Telford, George Burn   *NH 517 450*
over River Beauly   at the A9 junction

In Telford's original reports to the Parliamentary Commissioners in 1802 and 1803 he strongly pressed for a road to the north to facilitate commerce and so improve the lot of the inhabitants. Whereas Wade and Caulfeild had built roads to ease troop movements, and, in the final analysis, to hold down the population, Telford and his backers wanted to put roads where the local people needed them to transport the merchandise which would make Scotland more prosperous. The British Fishery Society was particularly active in promoting road making. The bridge over the Beauly and the one he built at Conon were, in Telford's words, 'the roots from which a great number of branches of roads are to proceed'.

His optimism was abundantly justified. In the following years not only did trade boom in fleeces, whisky, salt pork and fish, but farming spread in from the coastal strip and horsedrawn ploughs came into use as the whole northern country was opened up to the travel of ideas as well as of goods and people. In 1819 a coach already ran between Thurso and Inverness, and by 1828 Inverness had four coach building factories.

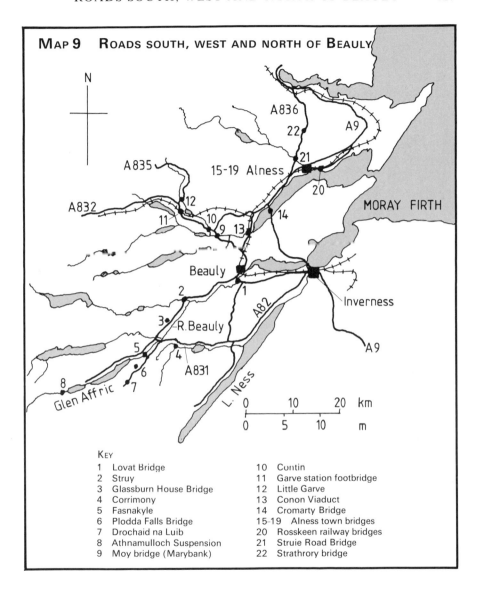

MAP 9   ROADS SOUTH, WEST AND NORTH OF BEAULY

KEY

1	Lovat Bridge	10	Contin
2	Struy	11	Garve station footbridge
3	Glassburn House Bridge	12	Little Garve
4	Corrimony	13	Conon Viaduct
5	Fasnakyle	14	Cromarty Bridge
6	Plodda Falls Bridge	15-19	Alness town bridges
7	Drochaid na Luib	20	Rosskeen railway bridges
8	Athnamulloch Suspension	21	Struie Road Bridge
9	Moy bridge (Marybank)	22	Strathrory bridge

79   Lovat Bridge over the Beauly River

This bridge was first planned as a wooden bridge of American pitchpine but this was not feasible unless a repairing fund could be tied in with it and this was not forthcoming from the government. (The repair and renovation of bridges was a concern to the Commissioners in the later years of the scheme when it threatened to swallow all the funds.) The stone bridge eventually built was damaged by the 1829 floods and had to be partly rebuilt by Mitchell and has received periodic attention ever since. The road has been raised at each end to flatten the hump somewhat but what you see is essentially Telford's bridge.

It is the largest one in our area with five spans (40, 50, 60, 50, 40 feet) and 470 feet of waterway, built of dressed sandstone of a particularly pleasant colour. It has semi-hexagonal piers which become refuges at parapet level except for the end ones which are solid topped. The parapet is 32 inches high and has a curved coping stone. There is a double string course, and each arch is outlined with an archivolt which continues across the cutwater tops. These are triangular and faced with thin metal strips. The voussoirs are almost square, and the spandrels not large for such a solid bridge over a river which can be violent. The banks of soft silty soil, blooming with bluebells, butterbur and wood anemones in April, show how the river floods, as does the rubbish caught on low branches and looking like old hair.

While it was being built the bridge was threatened by the passage downstream of large quantities of timber. Burns, the contractor, was using a centering described as thin as 'spectacle rims', and not likely to stand up to a battering from floating logs. The timber merchant, one Chisholm, was not disposed to be helpful. He said he could not use rafts (as a recent Act of Parliament required) because of the falls upriver. The crisis became so acute that sermons were delivered on it; but in the end Chisholm instructed his men to guide the logs past the half-constructed bridge with poles.

The bridge today has a crisp, almost new, look which comes largely from the clean colour of the stone and the neat articulation of the double string course. Southey commented, in 1819, that it looked 'almost as well as balustrades' but he deplored that no money was allowed for ornamentation of the Parliamentary bridges. This one cost £8,800. Southey also makes the following interesting comment: 'I learnt in Spain to admire straight bridges; but Telford thinks there always ought to be some curve, that the rain water may run off, and because he would have the outline look like the segment of a larger circle, resting on the abutments.'

On the right bank you can see a shabby pillbox—part of our defences in the last war—and on the left bank the War Memorial stands on a medieval motte. Here is part of the span of local history.

Continue down the A831. The long handsome STRUY BRIDGE (*NH 404 397*) can be crossed in a car and missed, because of the angle at which the road approaches it and the way the banks are overhung by alder and ash trees. It is well worth stopping to see, being a small, more rural version of Lovat Bridge. The best view can be had from a short way up the Glen Farrar road. (Telford called this bridge Varrar Bridge.) He built it between 1809 and 1817 with three main segmental arches and two side ones (30, 36, 40, 36, 30 feet) of coursed rubble with dressed stone voussoirs, parapet and string course. The triangular cutwaters on the piers continue up the spandrels as thin pilasters. Locally this bridge is called the garnet bridge because it is built of a schist with garnet in it. Some will tell you that in certain lights the bridge glows a pinkish red but, as the stones are not polished, I doubt it. I find the mid-grey stone harmonises well with the landscape. Telford seems to have set it with especial care in the valley so that, like all the best bridges, it belongs, appearing to have grown naturally and one can hardly imagine it gone.

Beyond Struy the road follows the curves of this famous salmon river, famous also for floods and spates, now somewhat tamed by the numerous hydro-electric dams. At Glassburn (*NH 396 344*) in a private garden, but visible from the road, is a wooden bridge over a falling stream. This solid, serviceable GLASSBURN HOUSE BRIDGE was constructed in 1987 by the owner Donald Mack, who has supplied me with the following details:

Two wire gabions measuring 6′ × 3′6″ × 3′6″, each holding 4 tons of rock, sit on the rock bed of the burn. The supports are two H-girders, 22′ × 9″ × 2″ resting on larch 'shoes' sitting on top of the gabions and wired to them. The deck is a larch platform 20′ × 3′6″ comprising 33 cross beams of 5½″ and 2″ secured and spaced by edging strips which also support the handrail which is 2′9″ high. Triangular supports from projecting crossbeams maintain the rigidity of the handrail.

The larch wood used was grown in the garden and milled at Struy. This bridge is a very nice example of the 'home-made' bridges I have seen crossing burns all over the Highlands. There is a useful book which can help with pattern designs if you are interested.

At Cannich you may like to make a detour down Glen Urquhart (on the A831) to Corrimony where there is one of the oldest bridges in this area.

CORRIMONY BRIDGE over the River Enrick (*NH 394 301*) was built in 1770 and is typical of the small arched bridges which were once common but now are fast disappearing due to the demands made by wider, heavier vehicles. This bridge takes the minor road up the short run to Corrimony where now little happens beside farming, but in prehistoric times it was an important religious site.

The river here flows over gravel where, in November, salmon spawn. Rather boggy grassland makes the banks and there is a long outlook

80   Glassburn House Bridge, a family project in the garden

81   Fasnakyle Bridge over the River Glass

towards the mountains of Glen Cannich. It cannot have been easy to site a stone bridge in such a wet place, and it is much buttressed. It is a single humped span of 30 feet built of rough stone with three buttresses on each abutment. The parapets are low and curve away at the ends, and are rather spoiled by modern rail and concrete posts. There is a plaque on the upstream side giving the date, and with some words I cannot read.

Return to Cannich and take the minor road to Tomich. Where this crosses the River Glass is FASNAKYLE BRIDGE (*NH 320 294*) set on rocks that slant up at about 45 degrees above water flowing fast and deep. Birches, white-trunked and lichen-draped, clothe the gully. The bridge springs up in a high, satisfying curve. It is calculated by the road engineers that the bed of the river at this point is a trough somewhat steeper and narrower than the arch above—like a reflection of the

82　　Plodda Bridge, on the fall's lip

bridge. If you can manage to climb down the mossed rock slabs you will have from below a fine sense of the power and balance of the arch.

It is dated in the early nineteenth century, its engineer unknown. Built of rubble with dressed stone voussoirs and parapet which is low and neatly curved off, except at one end where it sweeps round to make the roadside wall. The span is 29 feet and the overall length 14 yards. In 1967 it was repaired and strengthened. The arch was stripped down and then filled with concrete above a stone bottoming and a 2 inch layer of bitmac.

Near here is the large Fasnakyle Power Station which is built of gold sandstone from Burghead and decorated with pictish bulls. Sometimes the doors are open and it is possible to look in at the turbines. Both the road to Loch Affric and to Tomich should be explored to see varied Highland scenery and bridges. Tomich first.

PLODDA FALLS BRIDGE　　1890?　　unknown engineer　　*NH 276 238*
over Allt na Boadachan　　　　　　　　　　　　　　　　path from Forestry
　　　　　　　　　　　　　　　　　　　　　　　　　　　carpark beyond
　　　　　　　　　　　　　　　　　　　　　　　　　　　Guisachan house
　　　　　　　　　　　　　　　　　　　　　　　　　　　ruins

This charming little iron bridge is poised over the lip of the 100 foot Plodda Falls with an amazing view sheer down the jet of water to the River Abhain Deabhag which, at this point, makes a sharp S bend round a dramatic prow of rock. There could hardly be a more romantic spot from which to survey picturesque scenery and the owners of Guisachan House must have been delighted to have it in their grounds. Lord Tweedsmouth began in 1870 to lay out the paths through the woods and finally built the bridge, though its precise date is not known. It provided just the sort of thrill his jaded visitors from London, Paris, New York and beyond expected of the Scottish Highlands. The plan was for guests to stroll through the woods and happen on the waterfall. It is possible to see, below on the rocks, the remains of one of the first paths, or rather the hanging railings which edged it. Judging by the way modern visitors, in trainers and trousers, pant up the new paths to the bridge, I do not suppose that the Edwardian ladies, in their high heels and bustles, came here very often.

The bridge has recently been restored but is unaltered. It is of wrought iron arranged in a diamond pattern lattice with a flower at each intersection. Four curved triangular pieces hang down as extra decoration. The weight is borne by iron girders which rest on stone abutments built into the rock at the fall's lip. The railing, which is 50 inches high—and it needs to be a good height with that drop below— is supported by metal struts. The iron work splays round at one end to make an entry and at the other is closed by a gate. It is 15 yards long.

When my family and I first came to Plodda the bridge had not been repaired and the Forestry paths not constructed. We slithered, on our bottoms mostly, down to the river. In the winter of 1981–2, which was very cold, we found the falls cased in an ice tube, and the rock faces all coated in glistening ice, inches thick. If the water is low you can cross the ford near the carpark and then climb the bluff for a perfect view of the bridge.

Guisachan House, now a ruin, was inhabited until after the last war and one can just make out some of its features, notably the long ballroom at the back. In Lord Tweedsmouth's day it was kept in immaculate order, and the driveway continuously raked to obliterate wheel and hoof marks. The mighty wellingtonias, such a landmark today, would have been only moderately tall then. In the grounds was a dairy, a laundry, a house for storing game, and a small hydro-electric plant to supply the house. The remains of this can be seen below another, smaller waterfall.

On the Forestry road from Tomich to Cougie is the old patched bridge of DROCHAID NA LUIB over the Allt Riabhach (*NH 254 215*), by-passed now by a fresh stretch of track and a concrete bridge but, for me, it has great charm with its arches low to the water and its narrow length. Before any bridge was built, this was a fording place on the drove road from Loch Affric and points west.

83   Drochaid na Luib. Anderson shelters?

Drochaid na Luib is 18 yards long and only 11 feet wide. It has six small arches each lined with corrugated iron painted a rust red, and each set on rock in the stream bed. At first there was no parapet but latterly a guard of wire netting was put up. In the small splay of four of the arches there is a pipe to help the flow of spate water. The central pier has no such drainage hole and has been buttressed with timbers which, on the upstream side, are swathed in grass and sticks brought down by the water. The abutments are of stone and one has badly crumbled.

The date of this bridge is problematical. One Forestry engineer suggested about 1910. Another thought that the arches are really old Anderson shelters pressed into this odd use, which would give a post 1939 date. Perhaps both could be true, the shelters being used to prop up older spans.

This valley is one of the few places remaining where the old Caledonian pines grow naturally, though at one time they densely covered northern Scotland. Twelve years ago this bridge was still in use; we picnicked here and the children played in the water. Recently I saw it on a cold, sunny January day, and the water of the pond nearby, formed when the new road was banked up, was covered with soggy ice, cloudy and dimpled. Old rush clumps poked through, creating patches of clear water that reflected woods, hill, and sky. To the west stands Sgurr na Lapaich.

This is not the only bridge in this remote stretch of country. You have already crossed GARVE BRIDGE (*NH 267 223*) built at a place where two rough streams meet. It is a metal girder bridge with a wooden deck and sturdy rails supported by a central pier in the shape of a truncated elliptical cone—a pre-eminently serviceable bridge. Between this and Drochaid na Luib the Forestry road crosses a low ridge. Nearly at the top there is a passing place on the left where the ground drops away and there is a solitary dead pine quite near. If you stop here and look down towards the Allt Riabhach, where three pines stand on the bank, you will make out an old plank suspension bridge (*NH 264 219*).

It is not a bridge to cross lightly and, judging by the lichen growing thick as a crunchy carpet on it, has been disused for many years. It is a suspension of the simplest kind constructed by men who needed to cross the burn even when in spate. It has steel ropes twisted round tree trunks and supporting 3 feet wide wooden slats. By way of handhold there is a steel cable anchored to the deck by four wires on each side. There are stay cables anchored to posts in the ground but some have gone. It is a lovely place tucked under the craggy hill, shadowed by pines, with Clach Bheinn towering above and the grey-green lichen merging the bridge into its background so that it almost seems a natural product of the woods. When I was there it was a stag bounding down the hillside that led my eye to the place.

Return to Fasnakyle and take the more or less parallel road to Loch Affric, which is a superb drive down one of the most remarkable valleys in the Highlands.

ATHNAMULLOCH SUSPENSION	1920?	probably Alaistair Fraser and Rose Street Foundry	*NH 133 206* footpath from Affric towards Kintail

The way to this bridge is through unparalleled scenery and I would urge anyone who is up to a 12 mile walk to do it, quite apart from the interest of the little suspension bridge at the head of Loch Affric. The walk starts from the carpark.

The path circles the loch. The northern bank has perhaps the finest views down through the mountain valleys across Scotland. It is a grassy path cut by numerous burns and can be wet going, though the turf is wonderfully springy. You pass magnificent Caledonian pines in groups, pairs and singly, and a fine waterfall cascades from Coire Coulavie at Sputan Ban with a sound like distant sea breakers, and there are numerous small wooden bridges with lovely pools overhung by rowans, 'bead bonny' as the poem says. The southern path has a rough stone surface which I found rather troublesome but offers long

84   Athnamulloch Suspension Bridge at the west end of Loch Affric

sideways views into the Affric range, and, at the loch head, passes a beautiful curve of pale sand, a curve with a break in it formed by a jetty. The fluid shapes of nature are set off by a small red-roofed boat-house. Under the water you can see the tail of the sand spit fading away.

The sturdy suspension bridge crosses the River Affric just before it arrives at this meandering sandy reach and it looks very much as if the river has been artificially straightened at this point. The bridge is overlooked by a small white cottage with two sightless eyes and below are abandoned farm buildings. There is a ford downstream.

The date and history of this bridge are hard to come by. The Estate Office believes that it was erected around 1920 by the then factor Alaistair Fraser. An engineer friend of mine thinks that the Rose Street Foundry (now A1 Welders) used to supply landowners with cast iron suspension bridges like this. All the parts would be delivered and the

85   Athnamulloch, detail of rod stops

estate workers would erect it. However A1 Welders no longer have the records that would confirm this. It is a pleasant idea.

Certainly the bridge is well-constructed and seems set to withstand the weather for sometime yet, though, even in early September on a sunny day, the mild wind was singing quite a tune in the cables and rods. The pylons, which have small balls on top and a decorated joining bar, are 13 feet tall and the span is 90 feet. The cable dips to a minimum height of 3 feet 6 inches. The deck is of close set planks, 4.6 inches wide with a 4 feet walkway. It has a small rise and fall, and quivers and undulates slightly all the time. The clearance from the water is about 9 feet. The slender rods hung from the cable are fixed on with loops bound with wire. Beside each loop is a second one (unattached to a rod) which is there as a stop to prevent the rod sliding out of the vertical. These stoppers are placed on the side of the rod nearest the centre of the bridge towards which they might have a tendency to slip. On the lower cable, below the deck, the rods are attached in the same way but have no loop stops as here there is no risk of sideways movement. The wood beams of the deck are fixed to the lower cable with ring bolts. The abutments are stone; there is a wooden gate on the right bank, and both banks have splayed wood fence approaches. The cables are secured with large ring bolts into the ground behind and you can see the screwing sleeve, or turnbuckle, with threads going in opposite directions which tightens or loosens the cable. From here you can also see below you in the marshier ground a smaller, frailer suspension bridge which crosses the Allt na Ciche and is made of wood (*NH 129 203*).

86   Moy Bridge near Marybank over the River Conon, detail of cast iron piers

Athnamulloch Bridge stands below the great flank of Coire Crom. You look up the side valley that runs beneath Coire Donaich towards Sgurr nan Conbhairean, and westwards up the valley between Ben Attow and Sgurr nan Ceathreamhnam that leads through to Morvich on the west coast. In August scabious were flowering in the long riverside grass and our dog put up some snipe. On a previous visit in April the scene had been different, with snowy tops and white speckled flanks to the mountains with boiling grey clouds, but no less magnificent.

As you return you pass near Affric Lodge Bridge, an attractive timber trestle construction over the end of Loch Affric (*NH 186 229*). It cannot be visited without permission, but from the path you can see the long embankments, the trapezoid legs and the rise of the deck which is gated at one end. When reflected in the silvery water it is pretty, a miniature version of the road bridge at Broomhill, and of many other Highland bridges now swept away, such as those at Ballater and Boat of Garten.

2   *The Ullapool Road*

Going north west on the A832 the first bridge to see is at Marybank, MOY BRIDGE (*NH 482 547*) over the River Conon. It was built in 1894 by the Cleveland Bridge Engineering Company. Negotiating this long, narrow bridge, where two cars cannot pass, may distract you from looking at it but it is worth stopping to see. It is constructed of a mixture of cast and wrought irons. Cast iron takes compression well, so the piers are made of that metal; whereas the girders are wrought iron because that deals with tension better. It is 119 yards long and there are thirteen pairs of cast iron pillar piers linked by cross ties, and of these thirteen only ten stand in the river. On the upstream side are triangular projections to strengthen the structure against the rush of spate waters. There are stone abutments with short, decorated sandstone pillars. The wrought iron I-shaped side beams are topped by an iron mesh, probably of a later date.

To see CONTIN BRIDGE (*NH 454 566*) safely and in peace take the sign-posted right turn to Tonachuilty Forest. Here you can park at the end of the bridge which is a fine three-span one designed by Telford in 1816. It is built on a slope from west to east. The west abutments are set deep into the river bed, while those on the east are less so because here they stand on rock. The spans rise in height and increase in size, measuring 40, 45, 50 feet.

The line of this bridge can best be appreciated from the modern bridge which has superseded it. You will notice the triangular cutwaters which extend, in two stages, to the crown of the arch and then to the parapets, as very narrow pilasters. The parapet at 44 inches is unusually high; its ends are splayed and finished with square ends. There is no string course, although there is an impost, and the lack of this, along with the pilasters on the piers, accentuates the height of the bridge rather than its curve. Possibly this was done so as not to draw attention to the cant of the deck.

There is a charming cast iron footbridge at Garve Station (*NH 395 613*) which is part of the original Victorian work. It is a diamond mesh, painted cream with green rails. The steps go up to open piers with arched tops made of four metal pillars set in a square. If you are interested in railways I should see it soon, as the old gas lamps have already been whipped away. How inappropriate the modern lamps look on the bridge corners.

Until recently there was an interesting late-nineteenth-century wooden bridge over the Blackwater, up the road from the station, but it has been replaced by a metal bridge that looks efficient but out of place beside the old walled burial ground with its tall yews.

LITTLE GARVE BRIDGE spans the Blackwater about 2 miles west of Garve village (*NH 396 628*). It is signposted off the A836.

The rocks here are at a steep slanting angle and set with gorse, pines and larches. The brown water pours down with creamy foam and unwearying music and the bridge is poised here like an almost weight-less thing.

87   Garve railway footbridge. A particularly neat design

Its history is uncertain. It is definitely not a Wade bridge, though the local people will proudly tell you so; it may be a Caulfeild one and does remind me of other bridges of his. However, there is conflicting evidence as to whether a military road was ever built beyond Contin. Those interested should refer to books by William Taylor and by Haldane. Probably there was some sort of road between Contin and Poolewe, only intermittently kept in repair, and possibly the first part at least was built by the army. This bridge might be Caulfeild's and, if so, would be dated no later than 1767. After his day a Colonel Skene was in charge of the roadmaking and this may be his bridge. In that case the latest date would be 1790. It is an interesting sidelight on the military background to discover that Canadian soldiers, camped here during the last war, repaired the bridge.

The long sweep of grey stone goes up to the high hump and down again over one large arch of 50 feet and a smaller approach arch. The overall length is 250 feet. The parapet is 28 inches at the hump and has ends bending down in a quadrant. On the upstream side there has been considerable staying and repair work done. From the centre point you have a superb view of the wide valley, its farms and isolated buildings and the crags of Cnoc na l-Iolaire. When I was there in late April, skeins of geese were flying north-westwards, huge straggling Vs which constantly changed formation without losing shape. The two outer ends would drift into the centre, the leading birds drop back, a shimmer in the air and a new leader led on.

88   Skewed arches on the Conon Railway Viaduct

On the far bank of the river a lot of scrap metal has been dumped, mostly rusty rolls of netting and wire, and some bedsteads. This is deplorable in a place of such natural beauty. The other bridges on this westerly road are described in Chapter 10.

3   *The Road North*

CONON RAILWAY BRIDGE   1862   Joseph Mitchell   *NH 539 557*
over River Conon   west of A9 in
Conon Bridge
village

Mitchell himself called this bridge 'a great triumph of bridge engineering'. He put it across the river at an angle of 45 degrees, the north abutment being 304 feet lower than the south one. This skew means about 40 per cent more stone was used. (The bridge cost £11,391.) Usually skewing is achieved by placing of the piers but here they are parallel to the river's flow and each span is skewed, being built up of four separate arch rings, each placed a little ahead of its neighbour and each at right angles to the pier (see Figure 12). The distribution of forces in a skew arch are more complicated than in a square one and this method of Mitchell's was possibly a way of breaking it down into simpler square arches. It gives to each span a twisted effect, highlighted

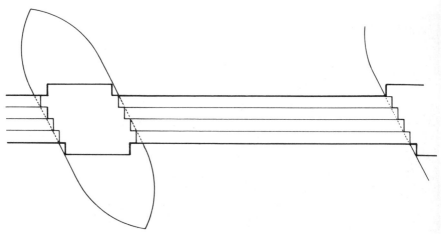

Fig 12   Conon railway bridge. The skewed ribs are set at right angles to the pier

now by the mineral deposits on the staggered ribs. When the river is low you can stand under the first arch on the right bank and inspect the stone joints closely. (You can walk to the bridge by taking the path along the bank from the children's playground near the pub.)

There are five segmental spans of 75 feet set on rectangular piers with oval bases. The matching abutments have the down curving walls typical of railway embankments. On the parapet is a plaque giving information about the local gentry who subscribed to the Inverness and Ross-shire Railway. They came from Gairloch, Ardross and Lochalsh, Lovat, Gadbol, Altyre, Aldourie, Raigmore, and London, and included the MP and the Provost of Inverness.

The modern road bridge here replaces one built by Telford in 1809 which was demolished because it could not take the weight of the construction traffic going north when the Dounreay nuclear reactor was built. It had spans of 45, 55, 65, 55, 45 feet. The toll house remains on the left bank. From here you can see plainly the different angles at which the bridges cross the river. It looks, from the remaining abutments, that Telford's was at right angles.

CROMARTY BRIDGE	1979 Crouch and Hogg	*NH 588 622*
over the Cromarty Firth	Fairburn Civil Engineering Ltd	on the A9 north of Dingwall

If you approach the Cromarty Firth from Inverness, and so see this bridge from above, the long curve is a harmonious focal point for the view of fertile agricultural country and the wide estuary. From this

89   Cromarty Bridge

distance the pairs of legs appear as low arches. A long embankment of grey stone projects from the south bank into the water.

At first the plan was to have much longer causeways on both sides, linked by only a short bridge. This would have restricted the flow of the tide in and out of Dingwall Bay to such an extent that ecologists feared its salinity would decrease which, in turn, might have affected migrating salmon and certainly would have disturbed the feeding patterns of the many birds who winter in the Bay. So a long bridge was designed, and the tidal waters' flow remains unimpeded.

The width to be crossed is virtually a mile and it is spanned by sixty-seven twin piers. These rest on piles driven 71 yards into clay. In all, 650 piles were sunk. Pre-stressed concrete beams support the deck which curves outwards to its full width of 8 yards. It has wide pavements and a railing of steel mesh and bars.

The two most noticeable features of Cromarty Bridge—the curve which leads the eye on across the water and the twin legs—enhance each other, for the constant apparent re-alignment of the piers as you move to one side or another, go away or approach, underlines the unmoving poise of the road's curve. The reflections of water on the piers and the underside of the deck only increase the effect of movement in stillness. This is on a bright day. Approaching from Dingwall on a misty morning, I have seen the legs fading into nothingness over the smoking water, and then this modern construction looked like something from a Chinese painting.

ALNESS TOWN BRIDGES

Telford stone arch	1810	*NH 655 696*
Pedestrian bridge		
Truss carrying a pipe		
Mitchell railway viaduct	1862	*NH 655 695*
Station footbridge	1975	*NH 660 694*

Alness is a small town with five bridges to look at. Four are grouped within a hundred yards of each other over the River Averon. The oldest is the sandstone road bridge, a single, segmental arch of 60 feet span, built about 1810 and designed by Telford. It is rather more decorated than was usual. Perhaps the citizens put some money into the project. It has a double string course outlining the gentle rise of the roadway and the spandrels have tapering buttresses which terminate as narrow pilasters, and are finished off, at the parapet, with rectangular blocks. Again the string course articulates these shapes. Originally the abutments were similarly decorated with tapered buttresses but, in all but one case, the left upstream side, these have been destroyed or mutilated by repairs and alterations to the road level. This is a great pity as the bridge is handsome and rather unusual. Recent repairs, such as renewing the voussoirs and a good deal of staying, have been carefully and unobtrusively done. It must have been some time ago that the abutment pilasters were dislocated and the botched repairs permitted. The bridge is solely for vehicles, and a footbridge has been placed beside it. This is a tube metal triangular truss with arch stabilisers, made by Tubewrights Ltd of Newport. The centre section is higher than the ends. It is different in shape, construction, materials and dimensions from the stone bridge, but is an honest and workmanlike thing and makes an interesting contrast. This would have been more apparent, and would have allowed the older bridge to show its looks better, had the tube bridge been a little farther up or downstream.

In fact, immediately downstream is the third bridge which carries a pipe. It is a thoroughly triangular bridge: its cross section is a triangle and its sides are triangular trusses.

Below this is the fourth bridge, the most imposing of the four, and yet easy to miss as you pass because it is screened by bushes on the bank and by trees growing on an island. It is a viaduct built in 1862 by Joseph Mitchell as part of the Inverness and Ross-shire Railway, and in his usual castle style which does blend with the countryside surprisingly well. It has two shallow segmental arches over the river and a small semi-circular arch over the road, now for pedestrians only. The courses of the soffits are skewed. It is of dressed stone of various styles including the very rustic. There is an impost and a chamfered edge to the parapet.

90   Alness footbridge over the railway line

The spandrel is ornamented with two towers with slits and square crenellated tops with dentils, and a semicircular arch between the towers topped by a coat of arms complete with motto. Altogether it is an elaborate bit of work. However, because the sun can never strike fully on this face and the corresponding one is masked by trees, Mitchell's details are for the most part lost and one is left with the pleasant effect of filtered sunlight falling through the little arch over the pier.

The fifth bridge is at the other end of the main street, a footbridge over the railway line but a very different thing from the cast iron footbridges remaining at some stations, and the ugly wood and metal spans used as substitutes elsewhere. This is an elegant bridge which neither wastes space nor cramps itself but carries the walker easily up, over and down in a smooth progress. It is built of re-inforced concrete and steel girders. The four spans are supported on five V-shaped piers. The airy upthrust of these piers contributes to the graceful appearance. The approaches are not stairs or ramps but a combination of the two— an easy slope with shallow, far apart steps. To bring a pram down here or a shopping trolley would not jolt the baby or the milk bottles too severely. The ramps curve round in a wide swirl so that they start and finish level with the first span. They are supported on columns of grey brick. The whole bridge is grey and white. The detailing is excellent. There is a thin white rail, a stone litter bin like a fullstop at the end of the curve and the light problem is solved by placing a very tall lamp-post, with two differently angled lights, in the centre of the circle made by the spiral ramp. The bridge was designed by Concrete Utilities of Ware in Hertfordshire. It is said to be 'a standard design' of the firm and was put here in 1975.

From Alness you can make a short run to Rosskeen along the Cromarty Firth. Take the B817 Invergordon road and, about a mile from where it reaches the shore, look for the railway underbridges. They are a short distance after the picnic site.

The ROSSKEEN UNDERBRIDGES (*NH 688 691*) were built in 1863 for the I and R Railway by Joseph Mitchell. All three do the same job of taking the line over small obstacles on a flat run of ground. Two are workmanlike, single segmental arches with variously shaped and angled wing walls to accommodate the burn and the road. The third was once over the entrance to the House of Rosskeen and was consequently embellished with curving walls, pillar-like abutments, a castellated parapet and even a coat of arms. It has, however, come down in the world. A nameboard now declares it is the entry to a Sewage Pumping Station.

Just behind the bridges is Rosskeen Church, its windows blanked with wood. It has an octagonal louvred tower, set on a square base with round windows and carrying a metal dome and vane. If this beautiful building were in an Italian village square it would feature on postcards and in guide books. Here it is almost forgotten. This place is a small corner full of historical contrasts. There is the church with its mounting block outside, the railway, beyond that the by-passed road bridge—a low, humped, sandstone arch that must once have spanned the Rosskeen burn where it entered the sea—then the steel giants waiting to be overhauled in the Firth where not many years ago the Fleet rode at anchor, and, as you return to Alness, you will see a prehistoric standing stone in a field facing the water.

The old north road, now the A836, continues over the high land to the north. The STRUIE ROAD BRIDGE (*NH 642 716*) is at the foot of a steep dip. Built by Telford in 1810 to span the River Averon it has been extensively repaired with brick in places. It is a single segmental arch of coursed rubble built where the road is on a ridge and narrowly confined as it crosses a basin-like hollow.

At the next dip in the road you find STRATHRORY BRIDGE (*NH 661 775*) a two arched sandstone bridge with a sloping road also built in 1810–11 by the Parliamentary Commissioners. Here the Strathrory river has cut a wide valley and left sharp banks on either side. Southey travelled this road in 1819 with Telford and described it as 'a noble display of skill and power exerted in the best manner for the most beneficial purpose', and said that the labourers were housed in tents donated by the army and, for their kitchen, had 'a hut with boughs'. As the road climbs beyond to the high point of 700 feet above the Dornoch Firth one is, even today, impressed by it; and all the more so when one considers the primitivee implements and the rough conditions of its makers. The two great bridges on this stretch, Easterfearn and Bonar, are dealt with in Chapter 10.

X

The Coast Road

Maps 10 & 11

1 *Dornoch Firth to Fleet Mound*

Telford's great north road runs down from the high point overlooking the Firth towards Bonar and one mile down on a tight bend it crosses EASTERFEARN BRIDGE (*NH 641 862*) over the Easterfearn burn. This delicate looking, old bridge can hardly accommodate the fury and number of the modern vehicles that power across it. Lorries with fish, whisky, builders' materials, food and much more cross it almost continuously. The greatest care is needed in stopping here. It is however worth the effort.

If you can manage to scramble down to the burnside, then you can fully appreciate the beauty and grace of the bridge. It is a semicircular arch sitting high in its buttressed abutments, and built of coursed rubble. The span is 40 feet. On the east side the wall continues into the road embankment and is an amazing dry stone construction formed of large rounded stones, somewhat bigger than footballs, chocked in with small flat pieces. In the top few courses the stones are a little smaller. It is approximately one-third the height of the bridge. Looking up you will see the drainage holes which are at gutter level. Some huge old birches grow in rocky footing close to the west side. The burn flows down from Ben nan Oighrean which towers above. Watching the flow of traffic from here you appreciate the builders' skill, the designer's nerve and the inherent strength of an arch.

BONAR BRIDGE over the Kyle of Sutherland	1973	Crouch and Hogg Tawse	*NH 609 915* on the A9

Whether you come to this bridge from south or north you cannot fail to be grateful to reach a crossing of the Firth. It is an obvious place for a bridge, but until 1811 travellers had to trust themselves to the Meikle

148

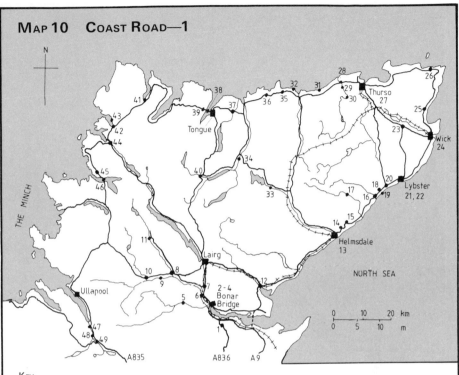

MAP 10   COAST ROAD—1

KEY

1	Easterfearn	18	Latheron Wheel Bridge	34	Estate bridges, Syre
2	Bonar Bridge	19	Latheron Wheel Harbour bridge	35	Strathy Bridge
3	Invercarron Viaduct	20	Latheron Bridge	36	Armadale
4	Ardgay Bridge	21	Lybster Bridge	37	Borgie
5	Wester Gruinards	22	Lybster Harbour bridge	38	Tongue Causeway
6	Shin Viaduct	23	Achingale	39	Achuvoldrach
7	Inveran	24	Wick Bridge	40	Altnaharra
8	Rosehall	25	Bridge of Wester	41	Drochaid Mor
9	Tuitean Suspension	26	Huna Mill bridges	42	Rhiconich
10	Oykell Bridge	27	Thurso Bridge	43	Achriesgill clapper
11	Estate bridges	28	Bridge of Forss	44	Laxford
12	Fleet Mound	29	Old Forss Bridge	45	Duartmore bridges
13	Helmsdale	30	Westfield	46	Kylesku
14	Ord of Caithness bridge	31	Fresgoe Bridge	47	Croftown
15	Ousdale	32	Dunes Bridge, Melvich	48	Auchindrean
16	Dunbeath	33	Estate Bridge, Loch Achnamoine	49	Corrieschalloch
17	Braemore				

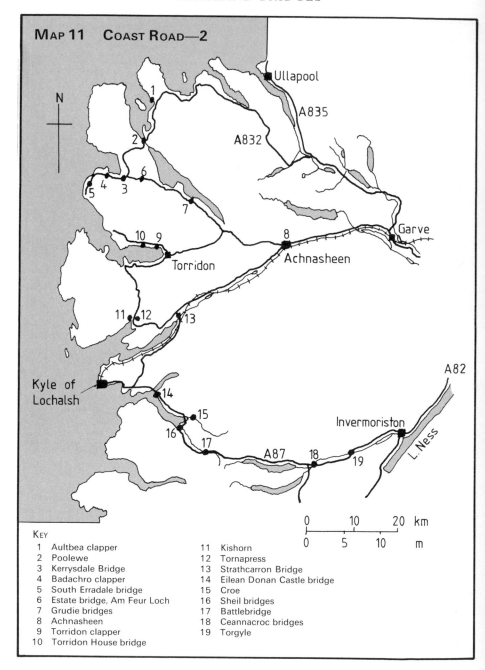

MAP 11   COAST ROAD—2

N

Ullapool

A835

A832

Garve

Achnasheen

Torridon

Kyle of
Lochalsh

Invermoriston

L. Ness

A82

A87

0   10   20   km
0   5   10   m

KEY
1   Aultbea clapper
2   Poolewe
3   Kerrysdale Bridge
4   Badachro clapper
5   South Erradale bridge
6   Estate bridge, Am Feur Loch
7   Grudie bridges
8   Achnasheen
9   Torridon clapper
10 Torridon House bridge

11 Kishorn
12 Tornapress
13 Strathcarron Bridge
14 Eilean Donan Castle bridge
15 Croe
16 Sheil bridges
17 Battlebridge
18 Ceannacroc bridges
19 Torgyle

91   Bonar Bridge reflected in the Kyle of Sutherland water

Ferry nearer the estuary mouth, or even go farther inland to the first good ford at Rosehall. The cattle were sometimes put to swim across here, but the currents drowned many. On 16 August 1809 the Meikle Ferry sank. It was loaded with people going to the Lammas Fair at Tain. Ninety-nine local people were drowned and an unknown number of strangers. Twelve were saved. It was this disaster that pushed ahead the plans to build a bridge at Bonar. The ferry, however, continued in business, run by a James McCraw from Golspie for an annual rent of £10. The mail coach used it rather than the bridge, presumably because it cut out two hills.

Telford's Bonar Bridge, 1812, was similar to the one at Craigellachie, and was described, by someone who spoke to Southey, as 'something like a spider's web in the air', and as 'the finest thing made by God or man'. There is a special poignancy in this comment because the speaker had lost his father in the ferry disaster. Telford intended two identical cast iron arches, but, because there was not the necessary rock base on the south side, he had to adapt his design and have two masonry spans and then an iron one of 150 feet. It bore assault from fir trunks embedded in pack ice, and from a drifting schooner crashing into it, but in January 1892 it fell, undermined by flood water. Another bridge was put in hand, with three iron girder spans of 70, 150, and 140 feet, and on a gradient of 1 in 50. Crouch and Hogg designed it and Arrol Ltd built it. To celebrate the opening one hundred guests had lunch in a local hotel. A speaker on that occasion predicted the bridge would be in use for 'centuries to come'. However it was replaced in 1973 by the bridge you see, also designed by Crouch and Hogg and built by William Tawse Ltd. All three Bonar bridges are shown on a memorial stone which also commemorates the names of the Commissioners who set the whole road building scheme in motion.

The present structure is a bowstring girder bridge, measuring 113 yards overall, and, like its predecessor, on a slight gradient. Below the War Memorial is a small garden from where a superb view of the bridge and the sea inlet can be had. Perhaps one longs for Telford's spider's web still to span these reaches of silvery water, but the modern bridge suits the scenery in its bare simplicity and strength. It entirely eschews the heaviness of much Victorian work. The road is hung on slender rods, and the upper arch is a wide diamond lattice of only seven sections. The railings are neat and unobtrusive and the stone abutments merge with the banks and make little visual impact, so that the eye concentrates on the splendid bow of steel. It is painted grey, except for the cream parapet below the railing and this gives definition at a distance. Here is an enormous landscape. The wide tidal river is the floor of the picture, silver and grey with tide current patterns of light and dark. From the water rise the hills, running out to sea and deep inland. If you are fortunate you will see salmon nets being shot from the haleing opposite. The peace of these reaches and low lying grasslands is summed up by the green-faced clock over The Carnegie Free Library stuck at 6.25.

From the tight neck at Bonar three chief roads radiate north and west. The purpose of this chapter is to follow the coast road north, then west and finally south and to take in a few inland diversions on the way.

The Carron Valley
The INVERCARRON VIADUCT (*NH 594 915*) is a fine bridge somewhat hard to locate. Take the minor road at Ardgay where the A9 makes a sharp bend. Having crossed Ardgay Bridge, turn right on the road to Oykel. Less than half a mile past the entry to Invercarron House there is a side road right over a small arched railway bridge. Take this and drive down to the farm where you can park. There is then a riverside path back up the river to the bridge. It has two tall segmental arches of dressed sandstone and a small arch on the right bank for flood relief, with rectangular voussoirs, rounded cutwaters and rustic ashlar stonework. It was built in 1867 by Joseph Mitchell and Murdoch Paterson his partner. There is a neat string course and a coping on the parapet to match. There is an echo here that bounces back and forth under the curve of the vault.

ARDGAY BRIDGE over the River Carron was built in 1818 by Alexander Thomas Muirson (*NH 585 910*). It is an unusually early bridge for a Highland road and reflects the importance of this area where many routes converged and there were substantial settlements, now mostly declined or disappeared entirely. The bridge was paid for by the second district of roads in Ross-shire, according to a partly obscured inscription on the bridge. It is a tall segmental span of 28 yards, strengthened with narrowly curved buttresses which connect with the abutment imposts. The curve can best be seen from the riverside path which is reached by steps in the left hand bank. Set quite high in the field is a small gated flood arch. You will notice modern voussoirs and staying rods, but doubtless Mr Muirson would be proud of his bridge, if he could see it being used by cars and lorries. The fisherman who was staring intently down into the water, dog at his side, has probably changed not at all.

WESTER GRUINARDS BRIDGE (*NH 503 927*) can be approached from either of the two parallel roads which run up beside River Carron. It is an iron lattice truss bridge of 65 feet span with stabilisers and set on stone abutments with wide approaches. It is built over a precipitous and pretty gorge in the river, where large exposures of white and grey metapsammite rocks jut into the water and make fine ledges for sitting to contemplate the bridge and the rushing water with its eddies of foam. The green-painted lattice fits well into the scene. It is 5 feet high, of the same width, and has a wooden deck. A small iron plate says that Brownlie and Murray of Glasgow were the makers. The date is possibly the 1890s.

At East Amat you pass a serviceable modern bridge and a two span arch rubble bridge over the Blackwater. Here the grassy banks, much cut into by water, are a contrast to the rocks at Gruinards.

The road ends at Croick church which is one of the Parliamentary Commission churches designed by Telford. Such churches were provided by the Government wherever a community was sufficiently large to warrant it. Telford built approximately forty churches and attendant manses. They are usually plain because funds were limited. Apart from a charge of £120 to the incumbent, the Government bore the cost. This church is simple but handsome. The large community it once served is now gone and the reason is the clearance of 1845. Inside the church you will find an account of this and the sufferings of the eighty people who sheltered in the churchyard and listened to their minister reading psalms. They were offered compensation of £18 if they agreed to go peaceably, but they were forced to go in any case, although none were in debt or want and all were prepared to buy their land for the market price—land that their ancestors had farmed for generations.

92   Shin Viaduct over the River Oykel, built by Joseph Mitchell

At this bare place, empty now of the quiet sounds of a rural community—the cattle, the dogs, the mill wheel, the carts, girls singing, children playing, the fiddler, the clank of pail and boots on ice on a winter's morning—one can imaginatively come close to those poor people sheltering under plaids in their churchyard.

A short way past the church, beside a stone building, a path runs down to a farm bridge where probably a crossing of some sort has stood for many years; and downstream is a simple wooden suspension bridge with timber pylons. From Croick several tracks run and many connect eventually with other valleys. These grand ways are only for experienced walkers. Those who take them will see a quantity of small footbridges for the use of those who walk the hills.

The Kyle of Sutherland

SHIN VIADUCT over River Oykel	1867	J Mitchell and Murdoch Paterson	*NH 578 953* over A836 at Invershin

Here, over the broad and peaceful Kyle of Sutherland, is Victorian iron work framing the view of a Victorian castle on a crag. It was part of the Sutherland Railway which Mitchell and his partner built for the Duke. A huge iron truss of 220 feet span, with nine complete box sections and two half sections, it has an elaborate diamond lattice side with an upper railing for the trackside and a lower one for an inspection walkway. The massive stone abutments have two semicircular approach arches on the south side and three on the north. It is 55 feet above the ordinary spring tides. It is one of the bridges Mitchell notes with pride in his *Reminiscences*. He says the railway was 'an intricate problem' with two arms of the sea to cross, not to mention mountains. Of this viaduct he mentions 'heavy rock-cuttings, breastwalls and embankments'. These are all in good repair today, although the old granite and sandstone station building is derelict. In 1870 the Culrain Station was opened immediately on the far side, and a half-penny ticket was sold in order to allow local people to cross the bridge. This concession lasted until 1917. Even so, some walked the line. The bridge was also the route by which the red squirrel, extinct in Sutherland from 1795, returned.

Joseph Mitchell's father John built the bridge at INVERAN over the River Shin in 1822 (*NH 575 975*). It is by-passed now, at the junction of A837 and B864, and not easy to see unless you scramble down to the riverside and even then it must be looked at through the new bridge. It is a picturesque hump of three segmental arches, built of coursed rubble with ashlar arch rings. The road is grassed over and not safe. Small pines, birches and sycamores grow out of the stonework. A photograph of 1928 shows it shadowed by an avenue of beeches.

ROSEHALL BRIDGE (*NC 472 023*) is of similar age, 1823, and style, but is still in use. It has two unequal spans with triangular cutwaters. It is built of rubble but the voussoirs and cutwaters are dressed. The parapet, unusually, is higher at the ends than the centre and curves neatly down to the road. There are stone steps down to the riverside and walks either down to the Kyle, or up towards some brochs. There are some old mine workings by the river, and the remains of a jetty. Up the hill the cottages which housed the Cornish miners are still known as the barracks.

Between Rosehall and Oykel Bridge on the A837 you pass, on the left, TUITEAN BRIDGE over the Oykel River. It was built in 1938 by John Henderson and Company of Aberdeen, but it seems as unused as some older bridges. It was the first over the Oykel above Bonar but has been superseded by a Bailey Bridge, just west of Invercassley near the old ford. When I arrived there, a lone ewe hurried before me over the bridge with an outraged waggle of her woolly hips.

It is a 300 feet span with steel girder pylons of lattice. The cables are wire rope (two above, and two below not reaching the centre) and the suspension rods are bolted to the cable. There is a wooden deck and railing of wire mesh. From the riverside it is clear that the deck's bow flattens in the middle section between the two lower cables. There is a

93   Glenmuick Suspension Bridge in good repair

local story that a car was once driven over this bridge. If so, the steps on the far side must be new. A battle once occurred here, when an ambush was sprung in the defile of the Tuitean burn which cuts down the hill above, thick with trees. To one side is a burial ground tidily boxed in a stone wall. When I saw the bridge on an autumn evening coolness was rising from the water, the sheep netting was hung with spiders on small webs and the leaping salmon made a rustling sound. The sun was catching the hilltops with crimson like a wine stain.

OYKEL BRIDGE (*NC 385 009*) by-passed by the A837 was built in the early nineteenth century. It spans the Oykel with a single segmental span of coursed rubble. The river comes pouring through the upslanting, almost vertical rocks on which the abutments are set. It is built in a deep hollow in the moor where a cluster of large trees, poplars, elms, a chestnut, a pine, break the bare stretches of hills. The arch

94   A second suspension bridge in Glenmuick by a deserted croft

frames a view of rocks and foaming water and itself is framed by hills, woods and heather slopes.

The GLENMUICK SUSPENSION BRIDGES (*NC 396 127*) are 9 miles up Glen Cassley which may seem a long drive on a narrow road but the bridges and their situation are worth the effort. Some way up the valley you will see that it narrows and forks. A little red-roofed house in the centre is where you are aiming. Just where a pine wood begins on the right, there is a broad track to the left. Park here and follow the track down to the River Cassley. Take care on the loose boulders of the steep path down to the bridge.

This unnamed estate bridge has steel ropes above and below, anchored in wooden rectangular piers by means of large steel eye bolts. The deck is lengthways planks over transverse beams and is covered with fine metal mesh to prevent slipping. Sloping wooden ramps rest on stone abutments. The sides are of sheep netting. The span is 50 feet, the width 39 inches, the cable height from the deck is 34 inches at the lowest point, and from the water 14 feet. Just above, the river is confined between rocks angled up at 45 degrees. Below is a large pool of brown water with suds of foam lazily turning. The rocks are metapsammite.

When I was there early on an October morning there was the thinnest ice on the larger puddles of the track and pads of frost between grass clumps, dewdrops heavy on fragile dead grasses, the white semicircles of cobwebs everywhere and the sun warm on my back.

95   Fleet Mound Bridge, the valve flaps and their pulleys over the arches

96   Fleet Bridge, west side, with new bridge going over in October 1988

To reach the second bridge, cross this one and follow the down streamside path towards the croft about a quarter of a mile away. Here the Muic enters the Cassley and the valley widens. The suspension bridge by the croft is in much less good repair, in fact my dog decided to wade through the stream. The construction is somewhat different. Here the upper cables are attached to wooden posts with wooden support stays, while the lower cables run through the posts and are fixed in the ground behind. The deck is wooden, the rails are merely two wires and you reach the bridge up suspended gangways. The stabilising cables from the bridge to the banks have mostly come away. The bridge is 98 feet long, 3 feet wide and the height above the water is 5 feet 6 inches approximately.

It is a beautiful, remote place. When I was there the stags were bellowing on the mountain slopes and a hind, startled by my dog, ran away up the brae, its powerful haunches working smoothly. I wondered whether the people who lived in this cottage above the ford ever regret leaving their old home.

The road north from Bonar Bridge skirts Dornoch and comes to Fleet.

THE FLEET MOUND BRIDGE	1814–16 Telford	*NH 774 982*
over River Fleet's mouth	Mitchell and Spence	north end of Mound, car-park provided

The earthen mound, or dyke, which Telford designed to keep the sea from flooding the low-lying land watered by the rivers Carnaig and Fleet, severely tested the engineers and the labourers. The work was superintended by John Mitchell, Telford's resident deputy in the Highlands. Although this man was almost without education, Telford saw his value and put immense trust and responsibility in his huge hands. (Telford himself had started life as a stone mason's assistant at Eskadale in the Borders.) For fourteen years Mitchell covered about 10,000 miles a year on foot and horseback, driving himself on through the bleak, harsh countryside and angry weather as fiercely as he drove his men, until he died in 1824. For the construction of the road north from Dingwall to Thurso his district superintendent was Thomas Spence. Until 1824 the men had only tents or improvised huts to protect them. One of Mitchell's last acts was the provision of better protection for them.

The bridge is the last section of the Mound and is now shadowed by the new A9 bridge which rises over it. Telford built only four arches; the other two were added in 1837 by Joseph Mitchell, John's son.

It is a bridge both beautiful and unusual, for its six arches are each closed by a wooden, non-return flap valve which prevents the salt sea water from flowing up river, while allowing the fresh river to flow out. Sometimes the flaps have to be opened, for instance to allow salmon up, and so they are fitted with chains and pulleys. Small stone buildings on each side house the winding gear. The effect of the arch curves, the black lines of the pulleys with their small wheels, the pale wood slats, the colours of the stone, water, seaweed, trees and sky is subtle and pleasing. It is also, of course, a remarkable feat of engineering.

When I visited it once in October 1988 the modern bridge was just taking shape above and the site was congested with machinery of all sorts and with various heated cabins and wash places for the men, which made a forcible contrast with the working conditions of Mitchell's men. In this yard were stacked bundles of concrete facing blocks with imitation stonework. I thought of the masons in 1814 patiently shaping stone by hand.

From 1900 to 1960 the Dornoch Light Railway, which hugged the coast, used the Mound and crossed the river on a viaduct, the piers of which can still be seen in the water. This railway was built by the then Duke who loved trains and drove his own. The posts upstream of the bridge are to protect it against floating objects, including ice floes. Loch Fleet is now a nature reserve of particular interest to ornithologists. On the information board in the carpark there is no mention of Telford, a quite extraordinary omission.

2   Fleet to Fresgoe

After the Fleet Mound, the A9 runs close to the coast under the escarpment of the inland plateau and gives long views of the Moray coast. As far as Helmsdale the railway follows the same narrow corridor. This line was not built by Mitchell, who had quarrelled with the Duke of Sutherland, but by his colleague Murdoch Paterson. The Duke wanted the line to run below the castle which Mitchell thought would ruin the outlook. He said the Duke was 'indifferent to refinements' and cared chiefly for yachting. However the railway was put behind the castle and a special station built at the castle gates. It is now a common-or-garden stop on the line, but ducal in decoration. There is also an impressive high viaduct at Golspie in the streamside park where there are also several small timber bridges of unusual construction. At Brora the road bridge is by Owen Williams. At Lothbeg the modern road has by-passed a Parliamentary bridge. Where the Loth burn enters the sea, through a cut in a low cliff, is a small railway underbridge which you can walk to along the shore from the caravan park.

97   Helmsdale Bridge, with ice house and war memorial

At Helmsdale fishing has been important for centuries. One of the main aims of the Parliamentary roads was to speed the despatch of fish. A bookseller called John Knox, who toured the Highlands in 1786 on behalf of the British Fishery Society, was one of the voices that helped to push the Government into action on roads. Before the days of refrigeration, fish was preserved by drying, salting or smoking; but the delicate flavour of salmon was thought to be hidden by these methods so, in the late eighteenth century, fishermen began to parboil salmon and then to pack it in ice for transport. For this they needed a store of ice, which was collected in winter and kept in underground chambers with arched roofs that look something like air-raid shelters.

It is therefore entirely appropriate that HELMSDALE BRIDGE (*ND 026 154*) should be beside the ice house. It was built in 1811 by Telford over the River Helmsdale, a high bridge with two segmental spans of 70 feet which are 25 feet above the water. It has stepped cutwaters that continue up the spandrels as semi-hexagonal pillars. The string course and parapet enhance its good looks, which are then somewhat marred by the modern lights set on short iron pillars. The illumination of nineteenth century bridges presents problems of scale and materials. One imagines that Telford, had he been asked to supply lighting, would have come up with something less fragile-looking. The bridge is of coursed rubble and stands with great strength over the river which drains a steep, bleak inland area spattered with lochs of all sizes. It was here that

98   Heavy traffic on the old Ord of Caithness Bridge

the short-lived Gold Rush of 1869 began, with a strike in the Kildonan Burn.

As well as the ice house, the bridge is overlooked by an impressive War Memorial, with a chiming clock, to forty men from the area who died in the First World War and sixteen in the Second World War. Downstream is the goodlooking modern bridge and below that the harbour. There is usually a coldish wind blowing at Helmsdale and bringing with it a smell of seaweed.

At the ORD OF CAITHNESS (*ND 052 182*) there is a sturdy bridge built about 1820 and set at a sharp V bend on the road, with a hill down on both sides. It continues to take the traffic which has grown steadily in weight and quantity since it was constructed, although it has been strengthened and repaired on the downstream side where the voussoirs are cement. Nevertheless it is a tribute to the engineer and masons who first drove this vital road north. If you walk up the valley behind for a short distance, you will see how the road comes down in a great U to the small arch. It is set directly on the rock and built of roughly coursed rubble, an almost semicircular arch with a square-topped parapet. The abutments curve round into wing walls which follow the line of the road downstream, that of the burn upstream. Those interested in the smaller Parliamentary bridges could look at a similar bridge, but only 10 feet in span, at *ND 047 174*. The spandrels of this bridge are slightly curved.

To reach OUSDALE BRIDGE (*ND 066 203*) as you drive north take a left turn and drive round to park near the bridge, which still carries a stretch

99   Ousdale Bridge, built by Telford

of the by-passed road. It would make a good place for a stop in a journey. The bridge is a tall, coursed rubble one with a narrow semi-circular arch, of 28 feet span and 40 feet high. It has triangular buttresses with squared tops and spans a burn where it runs in a gorge with wet boggy sides. Some buttresses appear in Telford's drawings for this bridge, but clearly others have been added and they can be picked out from the parapet as curved wedges. Obviously it was a difficult place to make a stable bridge. The abutments have deep foundations in the rock, foundations which are shown in the drawing continuing in steps below ground, down to the water level. From the streamside the arch romantically frames a typical view of the burn falling down low rock steps. In May primroses and celandines were blooming on the banks and thickly in the sheltered bays of the buttresses. A nesting bird flew out of the tree sprouts in the masonry. The modern road crosses on an embankment, with a tunnel to take the burn. Its total lack of visual appeal serves to enhance this tall, lonely arch.

The coastal road descends every few miles into a steep-sided valley. At Berriedale are two stone bridges built by Telford about 1815 and by-passed by two modern bridges with particularly unpleasant false concrete voussoirs. At Dunbeath a bridge of 60 feet span is in the process of being over-ridden by a new road. DUNBEATH BRIDGE (*ND 159 298*) was built in 1813 and is a typical Parliamentary bridge of medium size, with a low segmental arch and slightly battered abutments with dressed voussoirs. The harbour, with white-washed icehouse, is worth visiting. You will see a memorial to Neil Gunn who lived in Dunbeath as a boy. Two of his books—*Highland River* and *Morning Tide*—are the most truthful accounts of the old Highland life and spirit that I have read.

Do not leave this area without making a detour up the minor road inland to Braemore, which has a bridge situated in a remote and beautiful place under the conical Maiden Pap. BRAEMORE BRIDGE (*ND 073 304*) is built of coursed sandstone with dressed voussoirs and footing stones. The span is 36 feet. The road is 12 feet wide and is embanked as it approaches the bridge. Over the arch is a square drainage hole and the date of 1841. Although it has slightly battered abutments, which is typical of smaller Parliamentary bridges, this is not on the Telford road and was almost certainly built by a local landowner whose mason copied the common design. Beside this lovely bridge, a group of elms grow and in the near distance some pines stand on a small mound with a bare top. In May the lapwings, with their white and green plumage, were flying over the fields and a long-billed curlew stood in the old grass. Some ewes and lambs penned nearby made a constant noise, and the scent of gorse in the sun was strong.

The next harbour northwards on this road is Latheron Wheel and, if you turn sharp left over the modern bridge, you will come to LATHERON WHEEL BRIDGE (*ND 186 330*) built in approximately 1813. It has a span of about 40 feet and the road bends in a shallow U to cross it in a wooded valley.

The harbour itself (*ND 190 322*) is early nineteenth century and just above it is a bridge, now grass covered and partly ruined, which one might call LATHERON WHEEL HARBOUR BRIDGE, though the local people call it Wade's bridge. It used to carry the upper coastal road down to the harbour by an easier route than you have just driven. This can be traced up the cliff. At the top it runs straight ahead under a dip in the wall. The old bridge is a pretty bow placed on a bend in the Latheron Wheel river. It probably dates from the mid eighteenth century and has a single, segmental arch, about 48 feet in span, with a slight hump. The voussoirs are dressed stone, the rest rubble. The abutment courses are mortared but the spandrels are dry-stone or have lost their mortar. It is in a pleasant dell near the sea with wildflowers growing abundantly in August; and fishing nets hung up to dry on the cliff above. We were viewed by a seal who poked his whiskery snout up out of the sea.

Farther up the main road, at the junction of the A895, is LATHERON BRIDGE (*ND 199 335*), a small Parliamentary bridge not merely by-passed but swamped by vegetation in summer. It has a span of about 10 feet, but rises on tall abutments above the rock steps of a small unnamed burn. The parapet is only 2 feet tall with a flat top. Ferns, grasses, wildflowers, even a small fuschsia bush were growing on the roadway in August.

Just before entering the village of Lybster on the A9, you pass the old LYBSTER BRIDGE (*ND 242 359*) built in 1815. It is now ruinous, rabbits its only users. It is built on a curve with battered spandrels.

As you descend the steep road to Lybster Harbour, you will cross LYBSTER HARBOUR BRIDGE *(ND 243 351)*, and can observe the way it widens at each end to accommodate waiting traffic. Park on the quayside and then walk along the harbour edge to the footing of the bridge, a tall, semicircular arch over the Reisgill burn in a rocky gully. It is dated probably about 1849, but the engineer cannot be ascertained. It was built of coursed rubble, but half has been repaired with concrete. Altogether, it is rather an unusual looking bridge, due to its position, height and the line down the centre dividing old from newer. The Reisgill, after passing the bridge, is controlled by concrete banks and a low weir; then seaweed begins to appear as it enters the sheltered inner dock which is spanned by a footbridge. The pleasant harbour, dated 1849 but altered in 1882, has a red and white lighthouse, old fishery buildings and two slipways with stone bollards. In May the valley sides were yellow with massed primroses.

If you take a minor road about a mile north of Lybster village and signposted Watten, you will have an interesting drive across Caithness moor and pass the prehistoric burial mounds at Camster which look at home in the bare landscape. On the A882 near Watten in ACHINGALE BRIDGE (*ND 243 543*) which is a simple but handsome Telford bridge with three segmental spans, narrow voussoirs and triangular cutwaters protected with metal covers. It is placed on a curve on the road where it crosses a wide valley and a rushy field.

From here it is a short drive to Wick, passing Haster where there is another Parliamentary bridge. Wick, at the head of a long bite of bay, has always been an anchorage but did not have a harbour until Telford built one in 1824 for the British Fishery Society. He also built a bridge of three spans (48, 60, 48 feet) which is now gone. Those interested in the history of Wick, and particularly in the trade of 'silver darlings', cannot do better than visit the Heritage Centre in Bank Row which is housed in an old curing yard. The present WICK BRIDGE (*ND 362 509*) was built in 1877 by Murdoch Paterson. The contractor was Daniel Miller of Wick. It keeps the outline and size of Telford's. His was the second bridge here. The first was built in 1665. The present one is a simple, rather sombre structure seeming to mark the end of the fields and the start of the town. At the harbour is a more recent concrete bridge, also of three spans and with some decorative moulding.

The BRIDGE OF WESTER (*ND 331 587*), near Keiss and spanning the

Burn of Lyth, is a sad thing to see. Not only by-passed, but surrounded by the litter of the modern countryside—ugly fencing, ugly buildings, and, when I visited it, a pile of rotting rolls of straw. It was built in 1835 from a design Telford drew for Watten and did not use. It has two segmental arches, battered spandrels, and the parapet, slightly splayed, turning down through a quadrant. Inland extend fertile fields where cows were grazing in May, and beyond the river runs into the sea on Keiss beach, a beautiful stretch of sand.

Coming up the coastal A9 you reach the incomparable view from Warth Hill of 'the green Orkneys slumbering in the sea', then drop down to John O' Groats. Turn left at the junction with the A836 and, in less than half a mile at a sharp bend, are two interesting bridges of HUNA MILL (*ND 372 733*).

The smallest and oldest you will see at once, a small arch built of sandstone flags beside the old mill buildings. It is hard to come at the exact truth with regard to this bridge. It is said locally to have been built (not this time by Wade) but by Cromwell's troops, to 'give access'. It is true that the Commonwealth army came here in 1651 to deal with Montrose who had landed at the nearby Bay of Sannick. It is however hard to imagine an army troubling to bridge over a burn at a time when large rivers were often forded. (Cumberland's army forded the Findhorn at Fochabers in 1745.) Not only that, but it 'gives access' to nothing in particular. In all probability, before the days of good land drainage, this coastal shelf was very wet and the tracks did not run here but were either along the shore itself or on much higher ground to the south. The only possibility might be that bored and idle troops were set to build this as an exercise. More likely a miller with a taste for building made it for his own use and pleasure. It was recently put in order by the Manpower Services Commission.

The site is an old one. A millstone recently excavated and on display is Viking. the present mills were built in 1860 and 1902. The larger one, the more recent, is still in use. Milling is done from November until the spring field work by the family who has been milling here for at least 200 years.

Whoever built the bridge, it is a charming thing to look at, with kingcups and primroses on the banks in spring, and it makes a fine group with the mill buildings, the wheel and the mill lade, and the sea behind. From the field you can see the second Huna bridge which carries the road. It was built in 1876 and opened by the then Prince of Wales, who was visiting the hotel. It is a long, straight-topped bridge with two low segmental arches with dressed voussoirs and a small triangular cutwater. Flanking these arches are two much smaller ones, almost tunnels. The stream falls over a rock shelf just below, then slips away between overhanging grasses.

THURSO BRIDGE (*ND 117 682*) was built in 1887 by MacBey and Gordon of Elgin, and the contractor was John Malcolm. It fits well into the town, which was described in 1735 as a neat little town and still seems so. It has something of the quiet, order and good building of a provincial French town. The rather sombre bridge over the wide, shallow River Thurso has four segmental spans set on slender piers and one small flood arch. It is modestly decorated with a string course, keystones, imposts and a shapely parapet edge. The cutwaters are semi-octagons. Like the Helmsdale Bridge the lights are on tapered lattice pillars which strike me as unsuited to the design. The cost of this bridge was shared between the Country Road Trustees and the Police Commissioners.

Along the A836, travelling west, you come to the BRIDGE OF FORSS (*ND 037 687*) at a point where the river takes a sudden bend between high rocks, and there are two mills. It is picturesque, but rather dark and damp when I visited it, with a weir below the bridge. It is of coursed rubble with two low segmental spans and high spandrel walls. It is somewhat decayed. The millhouse is now empty. I was told by a local woman that it is very damp and her cousin 'suffered terribly' when living in it.

There is an older bridge at Forss, not marked on the OS map. If you return to the minor crossroads just east of the mill bridge and take the turning that goes inland, you will find that the road climbs to a sharp bend and then flattens out. On the left there is a wood. Stop where it begins and you will see below you OLD FORSS BRIDGE (*ND 041 675*)—a fragile arch over the Forss Water. A steep farm track leads down to it. It is quite difficult to get to the bridge because of gorse thickets but a way can be found.

It spans the river on a bend. Only the arch ring of the hump-backed bridge remains, demonstrating the strength of the voussoirs. These are formed of thin flagstones and the span is approximately 20 feet. You can walk over—despite the flagstones set on end to keep stock out—and see, through the thin grass covering, its structure. The footing is on the shelving rock of the river banks. Some cement repairs have been done here; otherwise no mortar is apparent.

Even though I visited it on a cold May day, it seemed an idyllic place. Marsh marigolds were thick and the yellow flag leaves were spearing up on the banks promising a great show in June. I saw and heard lapwings, snipe, oyster catchers and curlews. An old lady to whom I spoke in Reay, told me that her grandfather courted his wife on this bridge 130 years ago when it was an important part of the people's lives. She called it the Humpy Bridge.

Could this have been the line of the original coast road keeping to the

100   Old Forss Bridge, not on the OS map but still there over river

higher land? There is a stone dyke running uphill from the bridge and keeping south of the hill crest, and a minor road to Blackheath, to the east, is roughly in line with the wall. However, on the 1870 Ordnance Survey map, there is no road beyond the bridge. This, unlike the bridge at Huna Mill, is clearly a crossing of substance. It cost someone money and labour, and was needed.

Once on this minor road you may continue to the bridge at WESTFIELD (*ND 056 642*) along a road which gives fine views of a fertile Caithness farming valley, the river snaking in the bottom. The bridge has two unequal arches and the Forss Water flows under the larger. The triangular cutwaters extend up the pier. To the east the abutments are on rock; to the west on a sandy bank and so have a strengthening wall. In the vicinity are various cairns, a broch and standing stones around Loch Calder.

Returning to the coast road you pass REAY BRIDGE, another early nineteenth century bridge this one having almost semicircular arches. On the coast itself is a very small bridge, well worth a visit if only for the fine cliff-top walk, a brisk fifteen minutes. Take the road in Reay marked Sandside Beach and park at the harbour. Follow the track behind the cottages and beside a stone wall, through a gate and then uphill. The wall you are following is built of boulders and flags and is particularly fine. Where it turns a sharp left corner, you keep straight on over the short turf of the cliff top. You will cross the remains of a fallen wall and come to a wooden gate. Below you is FRESGOE BRIDGE

(*NC 951 661*), or perhaps it should be called Sandside Head Bridge. It is built over an unnamed stream just before it plunges down a cliff ravine into the sea. It is only 5 feet 10 inches high from the bed of the stream, and has a span of 7 feet 4 inches. It is partly ruined. Presumably, since no-one needs a bridge to cross so small a burn, it was built as an interesting project, perhaps by the local landowner. A farmer and fisherman, to whom I spoke, said the water of the stream had been altered and its flow lessened, but, even so, a bridge was never a necessity here. It is an extraordinary thing to find so near the cliff edge. You can return round the cliffs, which is a longer walk but beautiful. In May the lapwings were nesting on the bare top. Up here there is no sign of man until you turn the corner and see Dounreay's domes and its steam clouding over the sea. The little harbour (built 1830) is rather fine, especially the outer wall made of dry stone flags vertically set.

3   *Melvich to Kylesku*

The DUNES BRIDGE at Melvich (*NC 888 648*) is right on the beach and there spans the Halladale River. Opposite the Post Office a path leads down. Formerly, the river here was spanned by a suspension bridge, dated around 1900, which was erected by the landowner for the convenience of his employees, in particular the salmon fishers. The remains of this bridge—the bases of the towers and some metal eyelets and cabling—can be found in the dune sand. The present bridge was built in 1987 by the Third Troop of the Royal Engineers 48 Field Squadron, and as befits a military construction, is severely functional. Yet, deep in the gully of orangey-yellow sand with the dramatic swirl of river snaking through, it makes a fine picture as you come down the path.

It is a four span bridge, built of iron girders measuring 16 inches by 7 inches, and supported on three circular concrete piers with rectangular tops, and on substantial concrete abutments set deep into the ground with steps on one side and a ramp on the other. The entire length is 71 yards. The deck and railings are wooden and bolted to the girders. This can be inspected if you walk underneath on the pebbled sand at low tide. From this low vantage point you see the entire construction method, including the metal ties between the girders. It is about 8 feet above the mid-tide water level.

Menaced by the sea, as well as by the river in spate eating its way into the sandy banks, it is obviously a hazardous place for a bridge. Just below there is a breach in the dunes where a storm once blew through, perhaps the one that wrecked the suspension bridge. In August,

101   Loch Achnamoine Bridge

however, the scene was calm and the dunes warm. Knapweed grew among the stalks of marram grass and a few plants of the pale mauve flowered sea rocket (*cakile maritima*).

From Melvich the A897 road south towards Kinbrace crosses several small stone bridges. At Kinbrace it bends north towards the coast again. On this stretch are several estate suspension bridges which those interested in this sort of bridge (particularly simple ones with wooden pylons) might like to find. The prettiest is at the top of Loch Achnamoine (*NC 808 324*) over the sandy outlet of a small River Helmsdale. As you walk towards it, the deck planking makes a dash pattern of light and dark which adds to the charm. The upper cables of steel wire hang from iron bar towers 11 feet high, set in concrete, and anchored through iron sleeves into the ground where they meet the lower cables strung below the deck. This is two planks set on cross bars and the rail is the briefest caging attached to thin wire suspenders. The bridge is stayed from the centre point with steel wires which form a cross and are anchored in the banks. It is in excellent repair which is more than can be said for a suspension bridge near Syre. Take the B873 less than a mile along the River Naver. The bridge is below the road level at *NC 691 429*. Its construction is similar, though the pylons are wooden posts 6 feet 3 inches tall with an approximate diameter of 13 inches. It has lost all but one stay and sports a notice about danger but seems safe enough. We crossed it, and also the two suspension bridges which span the river and link to an island about 1½ miles farther down the road at

102   Armadale Bridge, details showing 'batter and step' style

NC 683 407. Here the warning notices have been made unreadable by weather and time, but our dog had no worries in scampering across. The wood posts are set into solid concrete blocks. There is no railing and the cable at its lowest point is 4 feet from the deck of planks. The cables run through metal sleeves, or turnbuckles, with threads in opposite directions so that the cables can be tightened.

Between these two sites along the River Naver you pass a memorial stone to Donald Macleod who witnessed and grieved over the clearance of the village of Rossal in 1814. The site of the village is the grassy space in the woods opposite.

Continue the drive along the north coast road, the A836, after Melvich, and you come to the unusual STRATHY BRIDGE (*NC 836 652*) over the River Strathy built in the 1920s by Owen Williams. It is the shape of a bowstring girder, with ten vertical spars which are 9 feet 5 inches apart, but, instead of being made of metal, it is cast in reinforced concrete. There are two bridges like it on Rannoch Moor (see Chapter 11) but few others. It is a bold design, perhaps somewhat spoiled by the pedestrian walkways and iron railings pinned to the sides. However these are convenient, as a passing walker said to me, and safe. I visited it in May, in a rainstorm, when it and the landscape was fairly bleak. In August, with haymaking in progress on the hill, it was altogether pleasanter and only needs some fresh harling to be seen readily as the handsome bridge it is. I particularly like the way the bow dips and flattens like an archery bow.

103   Tongue Bridge. The scaffolding was for the use of painters

On this coastal road you pass many stone bridges of the Parliamentary type, mostly by-passed. The one at ARMADALE (*NC 796 639*) is a good example to look at. The typical features could be summed up as: a straight roadway and flat topped parapet; built of coursed rubble with dressed stone voussoirs; the parapet ends curving down through a quadrant; and the slight splay of the bridge entrances. Perhaps the chief distinguishing mark of this period of bridge building however, is the slight battering (inward slope) of the abutment walls, along with the plumb vertical spandrels, which results in a square recessed ledge at the spring of the arch. The size of this ledge, or step, is determined by the size of the voussoir stones. Armadale Bridge has all these features and also a flood arch on the north side, partly blocked. The parapet is 3 feet high and 18 inches wide, which again is fairly standard. You will see many bridges like this on Highland roads. Most were built as a result of the Parliamentary road building scheme or of the undertakings of local Trustees who followed the good work. Another good and larger example is the by-passed bridge at BORGIE (*NC 668 589*) with two segmental spans and triangular cutwaters, but it is marred by ugly breezeblock walls across the entry. Surely bollards would have sufficed. It makes a V with the modern truss bridge built on a skew with four spans and sloping stepped ends to the parapet. These already have cracks.

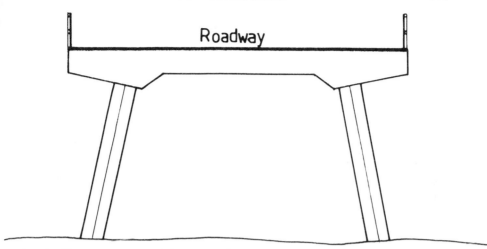

Fig 13   Tongue causeway bridge. The legs are diagonal to the road surface but
perpendicular to the underside

KYLE OF TONGUE BRIDGE   1971   Alexander Gibb   *NC 575 588*
over estuary   Alexander Sutherland Ltd   on A838

A bridge over this estuary was first suggested as long ago as the 1830s
by the local minister, inspired by the Fleet Mound. In 1971 this curved
embankment causeway and bridge was opened and much shortened the
previous drive. As you go along it you have superb views into the
mountains of Sutherland, notably Ben Loyal.

The bridge section at the west end is 201 yards long. It has to withstand the scouring rush of the tide and stands on seventeen pairs of stilt-
like legs fixed into moulded concrete beams. The deck is carried on reinforced concrete girders. The road is railed with metal bars and wire
netting.

My problem with this bridge is that, from most viewpoints, the pairs
of legs appear haphazard in arrangement. They are, in fact, set at an
angle to the deck, but are at right angles to the sloping underside of the
beam; see Figure 13. They do not, as at Kylesku and Cromarty, form
satisfying shapes, one against the other. The impression is more like a
pier, and perhaps this was in the mind of the engineer. It may be the case
that splayed legs take the waters inrush and outrush better than
perpendicular legs would do.

Just beyond the end of the bridge is a turning left, which was the old
road round the head of the Kyle, and here is a pretty Parliamentary
bridge dated about 1830 and called ACHUVOLDRACH (*NC 567 592*). It is
a simple, segmental arch, set on the rock sides of the steep, fast
Achuvoldrach burn, with a span of approximately 24 feet.

104   A bridge no more: a broken timber bridge at the head of Loch Loyal being
crossed by intrepid Thurso Scouts, August 1989

The A836 south of Tongue to Lairg offers a fine sweep of Highland
scenery running beneath Ben Loyal and besides its loch. There are, as
ever, frequent bridges over burns. The only one of real note is
ALTNAHARRA BRIDGE (*NC 568 357*) over the River Mudale where four
roads meet. An apparently remote spot, when one looks at the map, it
is in fact a busy crossing point and one cannot spend long on the bridge
in summer without having to move.

It has three segmental spans; the central one, which is the highest and
widest, is 24 feet. It is built of coursed rubble with substantial abutments
on the rocky banks and small triangular cutwaters. It has a slight hump
and splay-ended parapets, and the usual batter and step formation.
Unfortunately, the last section of the parapet on the right bank has
been replaced with a rail which destroys the symmetry of the bridge. On
the downstream side are some ring bolts. Their purpose is not clear to
me, as barriers against flood rubbish sweeping down are put up on the
upstream side.

Not infrequently in the Highlands a widening of the valley and a
slowing of a river before entry to a loch, as here, produces this sort of
quiet bowl in the hills. Despite its bareness and the winds that can bring
snow flurries even in spring and early autumn, it has a repose and a
beauty that refresh the spirit.

The A838 coast road continues to Durness and then turns south
down the Sutherland coast but running some way inland.

ACHRIESGILL CLAPPER   about 1700   *NC 257 541*
BRIDGE over   and later   on the B801 approx
Achriesgill Water   2 miles from Rhiconich

Clapper bridges are called primitive, but this should be taken to mean early, rather than lacking in technique. Anyone who has visited, for instance, Tarr Steps on Exmoor will know the high level of skill required to build stable, drystone piers and balance large flags across them. Many clapper bridges are very old, but in the Highlands, given the nature of the land and the force of the streams in spate, they are never much more than a few hundred years.

Achriesgill Bridge is probably getting on for 200 years old, but had clearly been altered before it was superseded by the new road which runs over the stream on an embankment with a tunnel for the water. It seems a pity that this had to be placed quite so close to the clapper, as it denies one a fair chance to look at it.

It has seven spans supported on square rubble piers, now mortared. The flags are visible as a layer between the top of the piers and the parapet wall. It would not originally have had this parapet. A good deal of infilling of the road has gone one, see Figure 14. The piers are 6 feet from the bed of the stream, as nearly as I could measure in pouring rain. The parapet is 3 feet high, and the flagstone 5 inches thick. Due to the road infill, the inner height of the parapet is less than a foot.

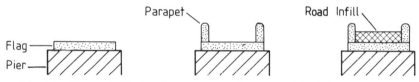

Fig 14   The probable development over the years of the Achriesgill clapper

The old bridge crosses the river at the best point in a steep valley and where the bed is shallow and pebbly. There has been revetting of the banks. It is worth expending a little imagination mentally to remove the modern road and houses and to see the bridge as a simple 'way continued' for long ago travellers, who might have been laden with peat creels, or fish baskets, some with ponies but most on foot with their dogs beside them. I see the dogs trot across the flags with tails aloft and ears cocked in the self-satisfied way of dogs accompanying their masters on business.

Other clappers can be seen at Aultbea (page 180), Badachro (page 181) and a particularly fine one at Torridon (page 184).

RHICONICH and **DROCHAID MOR** (*NC 254 521* and *360 604*) are two very similar bridges, both handsome and still in use. They have the typical vertical spandrels meeting battered abutments in a small step at the springing and the flat parapet with quadrant ends, though these have now been run into wing walls and embanking. The road at Drochaid has clearly been realigned. Both bridges were built about 1834 and both are in beautiful places, particularly Rhiconich at the head of Loch Inchard. Not far south of it, at *NC 246 514*, there is a miniature version of these bridges, now grass grown and partly ruined, but interesting to compare.

At LAXFORD (*NC 237 468*) there is yet another handsome bridge built about 1834, probably by the same team of engineers and masons. This is perhaps the best-looking of the three. It is on a fine stretch of river where the Laxford makes a huge S bend and the valley is wide. Notice the neat lipped drainage holes in the spandrels at road level. The abutments are battered but join smoothly into the spandrels without the step.

At the narrow inlet of Loch Duartmore there is a superb view downhill of the shallow green water and the modern roadway crossing an embankment. If you look inland you will see the old road bridge which also makes a pretty picture. It can be visited if you take the small road near the top of the hill on the south side, marked as a cul-de-sac. This is the old A894, and winds down to the bridge where there is now a fish farm complex (*NC 199 375*).

It is also a Parliamentary bridge dated about 1834 with the imprint of Telford on it.

KYLESKU BRIDGE over	1984	Ove Arup and Partners	*NC 227 338*
Loch a' Chairn Bhain		Morrison Construction Ltd	on the A894

It is the size of this bridge that impresses first, and then its simplicity, the unadorned geometry which suits the simple and majestic hills. Highland mountains do not usually have the toothed outlines of some European mountain ranges, such as the Dolomites. Wind and time have honed down the stone to massive curves, and in such a landscape this bridge is at home.

Built of pre-stressed and reinforced concrete, it is a five span bridge, the outer two spans on either side supported by the abutments and the legs, and the central span held by the extending cantilevers. It is a 300 yard long curve crossing a stretch of sea loch that is 142 yards wide. The bridge has a navigational clearance of 26 yards. Because of the strength of the winds here, which can reach 99 mph, and the remoteness, the bridge had to be strong and not to need much maintenance. For this reason there are no joints in the main part. The legs are hollow

105   Kylesku Bridge from below showing the shapes of the piers

structures and are canted inwards. The railing is in proportion and neat. It cost £5 million and part of this came from the EEC.

It is the piers which compel the most attention. They somewhat resemble those bent metal links that come in puzzles for the deft-fingered. If they were solid they would appear as triangular blocks with sloping sides, the outer ones wider than the inner. Because they are formed in this open, elegant way you see at least four separate, round-headed triangles that change constantly as you move, dividing and sub-dividing the space they contain. Their power, lightness and intricate geometry is best seen from down the bank where the curve of the deck and its deep plinth can also best be appreciated. There are viewing places on both banks. From the north you see the sweep of the roadway better; from the south you have a lower view and can walk round the hillside to see more. The ferry jetty is some $\frac{1}{4}$ mile inland.

Since criticism of the look of this bridge is sometimes voiced (though never of its usefulness), I asked the opinions of some people who had their cameras out. One declared it was in tune with the twenty-first century, another that it was elegant and in keeping, another thought it stark and should be painted green, and a fourth declared it was a perfect parabolic curve.

Above rise the rounded tops of Sail Ghom and Sail Gharb, with snow streaks in the gullies in May. Below is the greenish seawater, grey rocks and gulls crying. I think the bridge is a fitting successor, on this road that

106   Auchindrean, an unusual shaped truss

Telford first drove north, to his bold and technically innovative, yet plain designs.

At Ullapool the coastal road takes a long inland bend round the head of Loch Broom and along this stretch of the A835 are three interesting and related metal bridges.

The first is CROFTOWN BRIDGE (*NH 185 841*) over River Broom and just beside the main road. It is a light iron bridge with a single span built of three wrought iron ribs set on stone abutments that have angled sides. The railing is elegant, of two sizes of diamond mesh, and enhances the generally airy look of the little bridge, which leads down an avenue of limes and sycamores to the crofts on the far side of the Loch. It replaces a wooden bridge. Perhaps the slightly heavy stone abutments belong to that former structure. It has proved difficult to find accurate information on this bridge. It is possible, but not certain, that it was designed by Sir John Fowler in the 1860s.

AUCHINDREAN BRIDGE (*NH 196 805*), also over the River Broom, is opposite the entrance to the Leal Forest Garden, and is a structure said to be unique in Scotland. It was very probably designed by Sir John Fowler, who owned an estate here and was a famous civil engineer, in about 1865.

It is a fish- or lens-shaped iron truss, formed of two convex curves. The truss rests on granite abutments which extend into corbelled pillars which have a slice cut off their inner sides so that a slot can be made to hold the joined ends of the two curved girders. This slot is 4 feet from

107   Corrieschalloch Suspension Bridge built by Sir John Fowler, detail of cable sleeve over pylon

the deck and the height of the truss at the centre is 8 feet. The entire span is 102 feet 6 inches. The roadway, planking on iron girders, is hung from vertical lattice girders which are 10 feet apart. Each side wall is double and riveted together by the lattice road hangers, which are 1 foot wide, and by some cross bracing on the top. In form it is similar to one built by I K Brunel at Saltash. The whole thing is rusty and of a dark chestnut colour which, surprisingly, blends well enough with the grey-trunked trees and the russet bracken on the hillsides. Nevertheless it would be good to see it restored and painted, particularly since it is so unusual. A notice says it is unsafe for cars, but several passed while I was there. The rather scrappy wire netting, also rusty, further detracts from its appearance, but it still impresses as a handsome object curving away from its short, solid towers.

I can recommend a walk in the Forest Garden where there are some interesting trees, notably a young monkey puzzle growing as nature intended with branches to the ground.

The spectacular Corrieschalloch Gorge, which is 65 yards deep and almost a mile long, is spanned from ledge to ledge by a suspension bridge that combines an airy hammock-like curve with unobtrusive strength. CORRIESCHALLOCH BRIDGE (*NH 203 780*) definitely was built by Sir John Fowler, in 1867. He was one of the two engineers who designed the Forth Railway Bridge and there is more than a hint of that bridge's majestic power in this little structure, which was perhaps a sparetime diversion.

The span is 82 feet 6 inches and the abutments, well built into the gorge sides, are stone. The pylons are formed of two cast iron tubes that lean towards each other and join at a height of 8 feet 4 inches, making a narrow A-shape. The plane of this A is perpendicular to the span. Over the apex the cable passes in a curved sleeve at 9 feet high. The wire-rope cables come up from their anchorage points at an angle to the pylons and then, as they dip to deck levek, curve inward and the suspension rods are angled out to fit. It is this combination of diagonals on a dipping curve, along with the solidity and crispness of the pylons and the triangular truss carrying the deck, that gives this bridge its markedly buoyant looks. It is certainly not to be walked over merely to view the falls and the rock-clinging woods, beautiful as these are. Fowler's structure repays study and its characteristics reveal themselves as you look from one angle and another.

The bridge became the National Trust for Scotland's property in 1945. In 1977 cracks in the north side anchorages caused their replacement with concrete; and more recently the deck planking has been renewed and the mesh panels added. Otherwise it is as Fowler made it.

4 Aultbea to Achnasheen

AULTBEA CLAPPER BRIDGE   18–19th century   *NG 874 890*
over Allt Beithe   off A832 north of
                                                        Poolewe

Clapper, or flagstone, bridges such as this were common in the Highlands until recently. They are not only quicker to build than arched bridges, but also more readily repaired after the inevitable storm damage occurs. The technique is to construct dry stone piers of flattish stones basing them securely on rock in the stream bed and making them sufficiently close together that a single large flagstone can span from one to the next. The piers are long, narrow and parallel (more or less) and a series of flags of equal size are laid across the gaps to form a roadway. To this simple construction refinements are sometimes added, such as a further layer of stones and gravel to form a more durable roadway, and a parapet. Only in the hills will you find simple flag bridges. Most that remain have had an asphalt road put on them, also mortar between the stones, iron braces and other stiffeners.

The bridge at Aultbea has thankfully been retained and repaired. The new bridge, built in a clapper-like shape perhaps to harmonise with its older brother, hides the fact that the clapper was originally on the beach

108   Aultbea Clapper, near Poolewe

just where the Allt Beithe spilled over the pebbles to the sea. The position must have exposed it to some formidably severe weather. Even on a summer's day the grey houses and exposed hillsides hinted at bleak winds.

It is a sturdy bridge with rough stones placed in seven unequally sized arches, the final one on the right bank being at a skew and so longer than the others. The approximate size of a pier is 4 feet high, and 2 feet 6 inches wide. The spans approximately 3 feet and the length of each tunnel, that is the road width and the parapets, is about 15 feet. The 2 feet parapet is almost certainly a later addition. On the end someone has inserted a horse-shoe.

Over the River Ewe at *NG 858 808* stands POOLEWE BRIDGE a single span stone bridge with a stepped parapet going against the slope of the road. It is a quiet country bridge with views of the Torridon mountains. At the junction of the A832 and B8056 stands KERRYSDALE BRIDGE (*NG 822 729*) over the River Kerry built about 1840. The eastern abutment is placed firmly on a rock outcrop and the western abutment is at the end of an embankment carrying the road across a wide, somewhat marshy valley, which in summer is full of wild flowers. It is a pretty bridge spoiled by stay bars and a particularly ugly run of pipe which ruins the line of the arch.

Where the B8056 climbing from Badachro flattens out and there is a right turn to Aird, you should park beside the road to see BADACHRO CLAPPER (*NG 774 738*). A burn from Loch Bad na h-Achlaise runs under the road which is carried on a low clapper bridge. It can best be seen from the southern, or loch side for here no modern buttressing work is visible. There are three spans close together, the single 2 feet flags supported on dry stone piers about 2 feet high, and then a fourth span separated from the others by 6 yards or so. The first three run at right angles to the road but the fourth is on a skew and emerges 10 yards from the others. The crossway widths are 15 feet and 18 feet. These long low stone channels have drystone walls and are roofed, under the modern surface, with large flags, probably about 2 feet square. It is not a spectacular clapper (for that you must visit Torridon), but interesting in that it gives an impression of what many hundreds of Highland bridges must have been until quite recent times. A book published in the 1970s mentions such bridges all round the coast of Applecross. Most of these are now replaced with concrete and steel girders. If you can, in imagination, strip away the macadam and expose the solid squared flags spanning the drystone piers at no great height from the stream, and surely tufted with grasses and chance sown flowers, you will be seeing this place as it was until the road through to Redpoint was made. On the north side there has been some clumsy repairing with ugly breezeblocks, and a telephone cable has been hung across the water close to the bridge. This all seems a great pity.

SOUTH ERRADALE BRIDGE (*NG 744 713*), over the River Erradale on the B8056 near Opinan, is only half a mile from the sea and is best viewed from a distance up one of the straight roads running inland. As it is set low to the ground, it is easy to cross it unawares unless you are stopped, as we were, by slumbrous cattle unwilling to move off.

It is a long bridge of fourteen spans, each 4 feet wide and virtually semicircular, and supported on piers 4 feet high. It is built of dark stone and the parapet, with a rounded top and quadrant ends, has been harled. The piers are narrow and have triangular projecting edges on the upstream side, like simple cutwaters. The height of each arch is 68 inches from the floor—a word I use as the bed of the stream has been neatly covered with close-set stones which continue some distance on the downstream side. From the top of the arch to the coping is 40 inches, so that the total height of the bridge is approximately 9 feet and its length in the region of 100 feet. Although in August the river was calm the length of the bridge and the curved walls of the abutments, suggest that the water often pours across this flat valley full of reed beds. On the sea side it has been channelled into concrete walls and its rush broken by steps below the old paving.

On the road towards Achnasheen, A832, you pass a small wooden footbridge near Loch Bad an Sgalaig (*NG 857 720*). Look out for a corrugated hut painted white on the north side of the road. You can

109   Mountain bridge near Am Feur Loch on A832 approaching Talladale

park beside the hut. AM FUER LOCH MOUNTAIN BRIDGE was probably built by someone who wanted to try his hand at bridge building as there is strictly no more need of a crossing here than at many similar burns. It is charmingly simple, a light construction economically designed. The overall length is about 11 yards and this is spanned in two sections with a narrow central pier and abutments of stone. Two parallel tree trunks form each span and they are given a deck of narrow planks. On the road side there is a ramp up, and on the other two stone slabs provide steps down. The deck is 31 inches wide and the handrail is a single run of tough wire supported on iron posts. It is 39 inches above the water. It leads to the path up to Loch n h'Oidhche below the curiously shaped Baosbheinn.

At Talladale, a Parliamentary bridge built in 1843 has disappeared. Between here and Kinlochewe you pass the adjacent BRIDGES OF GRUDIE (*NG 967 677*) over the River Grudie. It is possible to turn on to the smaller bridge as it is now disused. It crosses the river at right angles, being set with great solidity on the strong pink rock walls of Torridon sandstone. It has a span of approximately 18 feet and rises from left to right. The other bridge is at a skew of about 45 degrees with a span of 36 feet. It is a reinforced concrete bridge with three ribs arching its vault and concrete abutments on to the rock but it is faced with ashlar masonry. The parapet is coped with somewhat aggressive rusticated stones.

These are not important bridges, but it is interesting to see how differently two engineers have approached the same crossing; and pleasant that the older one has been retained. From here a walk into Glen Grudie begins. Dominating the scene are the striated rock sides of Beinn Eighe—the name means File Mountain.

At Kinlochewe there was until 1985 a fine double span bridge, built in 1843, called Hunger Bridge because it was built as part of a relieving scheme at a time of famine, and also financed to some extent by Mary Hanbury, the mother of Osgood MacKenzie of Inverewe. In the carpark is a plaque with a picture of this vanished bridge beside its very ordinary replacement. Nearby is the War Memorial, which has a sword crossed with a rifle and a laurel wreath. There are the names of nine men who died in the First World War from Kinlochewe and of two who died sometime later of wounds; and then another two from the Second World War.

Various rivers and streams converge at Achnasheen, flowing down valleys with remarkable high-banked sides which were formed by ice. ACHNASHEEN BRIDGE (*NH 158 584*), over a stream flowing out of Loch a' Chroisg to join the Bran, was built about 1818 by Telford and is a typical medium-sized Parliamentary bridge. There was probably a ford here, and then a timber bridge. The segmental span is approximately 45 feet and set quite low to the river. The voussoirs are uneven and the parapet has a coping of irregular stones. From it you can see the lattice girder rail bridge over the Bran. It was built in 1870 for the Dingwall and Skye Railway and designed by Joseph Mitchell, who was the son of Telfords's deputy in the Highlands, John Mitchell. It has stone abutments and a single span, and is still in use for the trains to the Kyle of Lochalsh. The station retains its cast iron footbridge. Here the problem of lighting the stairs has been solved by placing a modern lamp standard close to each end. Those interested in railways might well follow the track down the A890 towards Lochcarron for there are several small rail bridges that cross the watercourses streaming down from the peaks above.

5   *Torridon to Torgyle*

TORRIDON CLAPPER   date unknown   *NG 888 567*
over Allt Ghoibhie   side road to Inveralligin

This beautifully constructed clapper bridge on the edge of the beach is one that should not be missed by anyone who appreciates craft and artistry. It spans the small burn which cascades down from Sgorr a' Chadail in a series of waterfalls and runs out to the sea there. You will

110   Torridon Clapper with its 'pavement' over the beach

find it on a bend in the road about half a mile from the jetty to the west of Torridon village, off the A896.

The bridge has two spans of 3 feet supported on a pier 46 inches high. The entire thing—abutments, pier, parapet and wing walls—is dry-stone, that is built without mortar and, in addition, the bed of the stream has been neatly paved before the bridge, through the runnels and even down on to the beach where it widens to an apron. The piers have triangular shaping on the upstream side. From the top of the span to the parapet coping is approximately 30 inches. The tunnels are roofed with flagstones close set and regular and these tunnels are wonderfully straight-sided. The wing walls however curve. Those on the beach side follow the bend of the land round; while those on the stream side splay out to collect water. Originally the flags, resting on the drystone piers, would have provided the road surface. Now this is covered over and an effort of imagination is needed to see this lovely bridge as it was, but you can easily inspect the workmanship and admire the balance and completeness of the whole. The colour of the stones, particularly under water, and the way the fronds of grasses and ferns enhance the economy and simplicity of the structure is another pleasure.

If you continue along this road there is a short pretty walk along a footpath beside the shore. It is signposted Inveralligin. This was the old road. It passes a jetty and a small house; just before the house is a single channel culvert, roofed with flag and opening on to the beach (*NG 876*

571). This takes the road over a tiny stream and is, in effect, a clam bridge—that is a clapper with only one span. A little farther on, through some trees, the road crosses the Abhain Coire Mhic Noboil on TORRIDON HOUSE BRIDGE (*NG 871 572*), which is painted green with ornamental posts at each end. It is a truss bridge with lattice work sides and stone abutments. It is 10 feet 6 inches wide, has stabilisers and a wooden deck. It is probably dated about the turn of the century and is certainly a contrast, in its Edwardian country estate style, to the clapper bridge and to the early-nineteenth-century stone arch which comes next. The river here, almost at the sea, has banks clothed in rhododendrons, gorse, rowan trees and beeches. Inland, there is a fine slanting waterfall.

KISHORN BRIDGE	mid 19th century	*NG 835 423*
over the River Kishorn		foot of road to
		Applecross, off A896

The wide, wet head-reaches of Loch Kishorn, through which the river runs, is a difficult place to provide a firm road, let alone a bridge. An embankment has been built where the rocky outcrops approach nearest; even so, it is a considerable gap to cross. Kishorn Bridge is hard against the western flank, one abutment securely on rock, and the other built against the end of the embankment, contained by a curving drystone wall.

Built of dark sandstone, probably quarried from close at hand, it is an excellent example of the later Parliamentary bridges which were put up in this area long after their original designer, Thomas Telford, had ceased working on them. One typical feature is the abutment walls which slope slightly inwards, or, one could say, are shallow buttresses. This batter coming against the vertical spandrels creates a squared step at the springing. The size of this step is determined by the voussoir thickness. This was the third main design Telford adopted for Highland roads, and it was continued by later masons. It gives a solid, well-proportioned look. Another feature is the way the flat-topped parapet bends down through a quadrant to make a sturdy curved join with the road. If you stand on the wide roadside above the bridge at Kishorn, all these features are plainly seen, and the way the mortared wing walls join the drystone retaining walls of the continuing road and its embankment. You can also see from here how the river has been revetted in low drystone walls, now partly grass covered, and led into the wider waters of the estuary.

The segmental span of this bridge is approximately 42 feet. It is built of coursed tooled rubble, the squarish stones fitting neatly enough

111   Strathcarron Bridge, with rubbish brought down by flood water

together. The arch is low to the water, the abutment footing having only four courses above the river.

Across the embanked road there is the smaller brother of this bridge at *NG 838 421* beside the modern road. TORNAPRESS BRIDGE over Allt Mor carried the A896 until not long ago but is now beginning to decay, with grass and even saplings growing between its stones. Again it is a Parliamentary type bridge with the usual features, but here, though the abutment is battered, there is no step visible since the soffit comes almost to the gravel bed of the river. Perhaps it has silted up. The road slopes uphill from north to south. The span is 27 feet. An unusual detail is the square topped pillar which finishes off the abutment.

| STRATHCARRON BRIDGE 1934 | F A MacDonald and | *NG 938 424* |
| over River Carron | Partners, Glasgow | on A890 near junction with A896 |

The flat valley at the head of Loch Carron is crossed by the embanked A890 which is taken over the river by this unusual reinforced concrete bridge built in the 1930s when engineers were still experimenting with concrete and, to some extent, using it decoratively as if it were stone; which gives Strathcarron Bridge a somewhat dated air reminiscent of Odeons and pre-war shop architecture which I find pleasing rather than otherwise. It is a more successful design than the same firm's concrete bridge at Dinnet on the Dee.

It has five spans each with three ribs on the shallow soffits. The piers are set on starlings with triangular ends, and they are narrow, rectangular blocks with two apertures cut in each. These are almost square holes with slightly domed tops. The cutwaters are round and extend above the piers. The parapet is rounded and has iron bars, painted green. Over each pier the run of bars is interrupted by a rectangular block marked with vertical grooving. The abutments are formed of a combination of the same grooved blocks and the round-nosed columns. The pier rises 9 feet from the river bed, and the aperture measures 4 feet by 4 feet 10 inches. The overall length is 72 yards. The detailing is satisfying and unobtrusive. For instance, in the centre of each span the railing supports are double.

Not far from here the road crosses Attadale Bridge, one of a series of County Council bridges between Torridon and Stromferry. They are serviceable but in no way distinguished. The one at Attadale is very untidily combined with a railway viaduct; the piers of the two bridges jut uncomfortably against each other.

The castle at Eilean Donan (*NG 882 259*) is known to people who have never visited the Highlands, for its romantic image has appeared on innumerable calendars and shortbread tins. The original castle was destroyed in 1719 as a result of shelling from the sea. Some Spanish troops had occupied it and were hoping to raise support for the Stuart cause. This castle had no bridge, and a photograph of 1879 shows nothing here by ivy clad ruins on a small isle offshore. In those days of no roads, people either reached the castle on horseback at low tide, or by boat. Before the nineteenth century the villages of the western seaboard used the sea as their road, almost all traffic, in goods and people, being by boat.

When the castle was rebuilt, in 1912–32, the bridge you see was provided and in keeping with the castle architecture. It has three main spans and a small flood, or tidal, arch. In 1985, when a film set in the 1700s was shot here, the 1930s bridge, garlanded with skeletons in chains, looked most convincingly ancient.

CROE BRIDGE (*NG 959 213*) is a fair-sized bridge of dark brown-grey stone low in the valley of the River Croe, inland of the modern road. Its three spans are 25, 28, 25 feet and it was designed by Telford and built about 1820. It has a string course and impost mouldings over the cut-waters to match. The string course tucks round the square end of the parapet and finishes it neatly. Though still in use, it is somewhat neglected with grass and saplings growing in the crevices. It is masked by riverside trees. This is a pity. If the banks were cleared of scrub, the handsome arches would be visible from the modern road near Shiel and would make a focal point for the valley head below the magnificent

112   Eilean Donan Castle and bridge with a snow storm just passing

hills. It would be a pity if such a handsome piece of engineering was allowed to lie forgotten and decaying in the alders and birches.

At Shiel there are two bridges to see. The more important and the obvious one is SHIEL BRIDGE (*NG 935 188*) over the River Shiel on the road to Glenelg just off the A87. It was built by Telford in 1820 (and called by him Sheal). It is a lovely, single-span, stone bridge perfectly suited to its mountain surroundings and echoing the simple lines of their lower slopes. The span is 65 feet. It has narrow voussoirs, a string course and a darker edging to the parapet—details which emphasise the slenderness of the bridge and its considerable hump. There is an unusual feature in that the abutments are decorated with a flat pillar-like structure with an upper rim that connects with the string course. I have only seen one other bridge with this detail.

Just beyond this bridge, in the direction of Glenelg, there is a left turn which leads to the second Shiel Bridge, built over the Allt Undalain. This was on the line of the military road which came down from Glen Moriston on this side of the river. Until 1967 the A87 main road ran this way, which route meant two awkward turns, before and after the Telford bridge. This may be the original military bridge, but is much more likely to be a later replacement. It is now judged unsafe and stayed. What might be a picturesque spot, with yellow flag irises growing in the boggy grass, has been allowed to become something of an agricultural dump upstream which is a pity. If you can struggle downstream to the river confluence just below, you will be rewarded (*NG 937 186*).

Half way up the long climb to Loch Cluanie through Glen Shiel you pass, on the left, another Telford bridge left behind by the new road at *NG 990 132*. You can easily turn off here and enjoy the superb lines of the low, solid arch at leisure. Built in 1814–17 it crosses the Allt Mhalagain, a single span hump bridge. Its masonry work is rather finer than usual. The spandrels are smooth dressed stone and the abutments rustic ashlar. There was a battle near here in 1719 when a Spanish army invading in support of the Stuarts was defeated by Government troops. For this reason I call the place Battlebridge.

The upper section of Glen Moriston, where the valley widens and the river loops through farm land and the larch and pine plantations on the lower hillsides, is particularly beautiful. On the day in April when I was there the river was the strong bright blue of sapphire and the fresh larch foliage was picked out, olive green, by shafts of sun. Sheep and lambs grazed in clusters. Primroses grew in little bunches. Yet snow storms hung on the mountains as if their flanks were smoking. It was in this flattish upper section that the military road (Caulfeild's not Wade's) crossed the Moriston having come over the tops from Fort Augustus. It was built between 1755 and 1770 by parties of soldiers and is still there to be walked. It gives superb views of the surrounding countryside and crosses several old bridges. The one over Moriston is gone.

OLD CEANNACROC BRIDGE (*NH 228 107*), just off the A887, was built by Telford in 1808–11 and is now partly ruined. The stone is a rather dark rubble, the voussoirs dressed. It has two unequal spans (50 feet and 36 feet), small triangular cutwaters and battered abutments. It is humped with a 15 feet roadway. The spandrels are now held with staying rods which somewhat spoil the looks of the bridge but on a sunny day down by the river it is a pleasant spot. A few birch saplings sprout from the masonry and the roadway is grassed.

It was near here, before this bridge was built, that Doctor Johnson and Boswell stayed at an inn on their way to Bernera in 1773. The hamlet of Aonach, now gone, had 'three huts one of which is distinguished by a chimney', and that was the inn, built of turf and thatched with heather. The inner walls were made of neatly plaited wicker.

From this bridge you have an interesting view of the new Ceannacroc Bridge, built in 1950, a short way upstream. It looks made of rectangles. This is partly because it is set skew to the river. There are darker lines on the road surface that show the angle of the piers underneath to be approximately 50 degrees. It is an iron girder bridged faced partly with concrete and partly with stone. The piers are triangular, like cutwaters extended to deck level, and, where they stick out below the parapet, have curious triangular cutwaters of their own, but on one side only,

113   Old Ceannacroc Bridge, seen from new road which by-passes it

upstream left side, downstream right. The parapet has stepped ends. This bridge, which superseded Telford's in 1950, will itself soon be by-passed by a new road coming up the glen. Building was due to start in August 1989.

TORGYLE BRIDGE over River Moriston	1808 and 1828	Telford and Joseph Mitchell	*NH 309 129* on A887 about half way between Cluanie and Invermoriston

This handsome, solid bridge of tooled rubble has an interesting history. Originally built by Telford between 1808–11, it was destroyed by floods in 1818 and rebuilt more or less to the same design by Joseph Mitchell in 1828. What broke Telford's bridge was not the rush of water itself but the battering of 4,000 birch timbers which had been piled on the bank upstream and were swept down. The use of water for moving timber was an old and jealously guarded right, and in an area of few roads and many hills a river was the obvious route to use. The men who practised this dangerous craft were called floaters. When stone bridge building began in earnest the hazards of log floating became apparent. Several bridges were brought down, notably Potarch in 1812, and others menaced, such as Ballater and Lovat. An Act of Parliament of 1813 forbade the floating between March and November under bridges

114   Torgyle Bridge over the River Moriston, a Mitchell rebuillding of a Telford
bridge

being built, except in the form of rafts. Torgyle was destroyed in
January—not the close season, as it were—and in fact the logs that did
the damage were not hefty timbers but thin spars for making herring
barrels. A temporary wooden bridge was put up by John Mitchell. The
workmen had to stand in 4 feet of freezing water. It was John's son
Joseph who re-erected the stone bridge.

It cannot in all respects be as Telford's was, for the first bridge had
spans of 38, 40, 38 feet and this one is 45, 50, 45 feet. It has triangular
cutwaters which continue up the spandrels as rounded pillars. It is
quietly but effectively decorated, firstly by the way the stones are
arranged with large blocks alternating with smaller ones set the
opposite way (a technique known as Aberdeen sets), and secondly by
the voussoirs which are alternate blocks of dark schist and lighter
granite, giving a striped effect. There is a string course and an unusual
open box decoration on the tops of the pillars, which also have slits in
them. From below it looks as if these were refuges, but this is not so. The
solid pillar tops add an 18 inch wide shelf to the parapets. These, on the
north side, have ends to match the pillars, but on the south they bend
round to form roadside walls. It is probable these details are Mitchell's,
for Telford, strictly limited in the money he could spend, rarely allowed
himself decoration. Southey called this 'farthing economy' on the
Government's part. The river banks are revetted for about 50 yards on
either side by drystone walls which run into the abutments.

115   Torgyle, detail of tower head

I particularly like Torgyle with its handsome proportions and colouring, and its restrained but telling decoration. I visited it last on a sharp day in April when a north east wind sent snow crumbs whirling past and puckered the shining water. The tops were white, birds were active on the banks, and three children on bicycles and in bright anoraks came whizzing over the bridge calling to each other.

XI

Argyll

Map 12

One way to approach Argyll is from Ballachulish on the A82 over Rannoch Moor, so long avoided by roadmakers because of its treacherous bogs. The military and drove roads kept to higher ground to the west and came down to the line of the present road near the modern chair lift. Near here is ETIVE BRIDGE (*NN 265 539*) built by Owen Williams in the 1920s and recently smartened with a coat of cream paint which lifts its arch away from the hillside. It is in effect a bowstring arch built not of metal but of reinforced concrete which must have been poured into the moulds on site. It is unusual to see a concrete bridge this shape. There is another at Strathy on the north coast (see page 171), also by Williams.

Embanked approaches bring the road on to the bridge. The abutments are substantial and are topped with pairs of rectangular blocks measuring 39 × 58 inches and 54 × 78 inches. The arch itself is 3 feet thick and rises with eight vertical beams. The sides are runs of neat slim concrete posts between the large beams. This I think is a better solution than that of pinning the paths on the outside as at Strathy. This is a solid, dignified bridge that suits the massive hills and the bare valleys of Rannoch. Its twin is nearby on the same road at *NN 313 444*, known as Tulla Bridge, on the way to Bridge of Orchy.

Caulfeild built a BRIDGE OF ORCHY (*NN 296 396*) in 1751 as part of the military road which went to Glencoe on the route now taken by the West Highland Way, round the head of Loch Tulla. This is probably a reconstruction of his bridge. It is a single segmental span with a hump, of rubble, with abutments set on the rocks and a slightly splayed entry. An unusual feature, possibly added recently, is the flag pavement. It is set in an extremely bleak section of the way north—a useful pass between formidable crests and used by the modern road, the old road, the railway and the electric pylons. The Wordsworths travelling through Scotland in 1804 called it 'a bare moorish waste'. They had spent the night at Inveroran which Dorothy saw as 'a flower in the

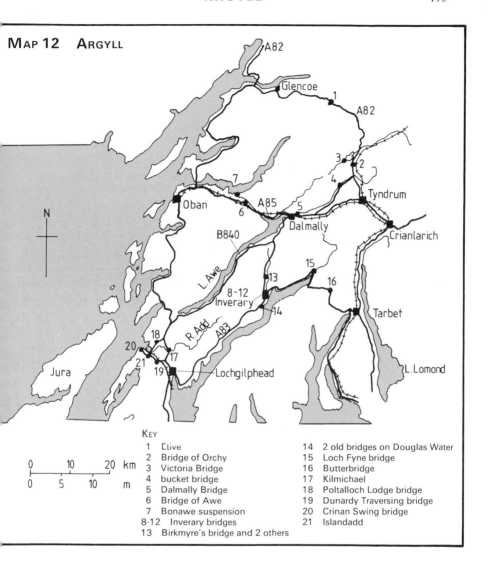

MAP 12   ARGYLL

KEY

1   Etive
2   Bridge of Orchy
3   Victoria Bridge
4   bucket bridge
5   Dalmally Bridge
6   Bridge of Awe
7   Bonawe suspension
8-12   Inverary bridges
13   Birkmyre's bridge and 2 others

14   2 old bridges on Douglas Water
15   Loch Fyne bridge
16   Butterbridge
17   Kilmichael
18   Poltalloch Lodge bridge
19   Dunardy Traversing bridge
20   Crinan Swing bridge
21   Islandadd

116   Etive Bridge on Rannoch Moor. An unusual use of concrete

desert'. It certainly has a softer air. You round the corner and enter a group of Caledonian pines through which the green verges and calm water of Loch Tulla are seen. Not surprising to learn this place was a stance on the drove road south, that is a resting place where men and cattle were fed, watered and rested. The road ends at VICTORIA BRIDGE (*NN 271 422*) which was built about the same time as Forest Lodge in 1843. It has two segmental arches, a low plinth pier and small cutwaters. The string course and the parapet are straight and the parapet is topped with rectangular blocks. In fact it is a typical example of a modest mid-nineteenth-century country bridge. It may, or may not, have connections with Queen Victoria who may, or may not, have come over it on her honeymoon tour of the Highlands. Legends abound and facts are few where this bridge is concerned. The old tracks run in several directions from here.

As at so many old sites it is hard to imagine the crowds and noise here when a herd of small, muscular black cattle were pastured. Easier perhaps to visualise William and Dorothy Wordsworth in their jaunting-car coming down from Glencoe to breakfast at the inn. Dorothy found in the inn kitchen 'seven or eight drovers, with as many dogs, sitting in a complete circle round a large peat fire in the middle of the floor, each with a mess of porridge, in a wooden vessel, on his knee'. Today in the same inn there is a notice to request that you do not remove your shoes and socks, as 'plastered feet' may offend other customers.

An interesting side road from Bridge of Orchy south is the B8074

117   A bucket bridge in Glen Orchy

down Glen Orchy towards Dalmally. About 2 miles along this road you
pass a pine plantation on your left, then a small bridge over a burn at
Invergaunan. Here look out for the **BUCKET BRIDGE** (*NN 276 367*) over
the River Orchy. A bucket bridge works on the basis of inertia and
pulleys. It is a large wooden box with two pulley wheels fixed on each
long side and it is suspended on these wheels from two steel cables
anchored into concrete on the banks. Then a rope loop is attached to the
box at each end and goes through the pulleys and is also attached to the
concrete blocks. You climb into the box which then shoots rapidly to
the centre of the cables over the water and a little way beyond with its
momentum. After that you have to haul yourself into the far bank. As
the box is heavy, even when empty, and resists the final few feet of its
journey, one would need at least two people to cross with any degree of
ease. Being alone I did not try—especially in view of my dog's probable
panic if I had disappeared in a box over a gulf of pounding waters she
had no hope of swimming.

    At Inverlochy turn right to Dalmally. The low-lying land of the
Strath of Orchy and the river-cut ground at the head of Loch Awe is an
area full of history, an ancient place of settlement, being at the crossing
of several ways. Long before the days of roads made for vehicles it was
known to travellers, to drovers and to marauding men. **DALMALLY
BRIDGE** (*NN 167 276*) is also known as Bridge of Urchay. It crosses the
Orchy and was built in 1780, but it must have had predecessors, and
possibly there was a ford and a ferry hereabouts also. It was constructed

directly after Awe Bridge and by the same contractor who thus economised on timber for the centering. It was designed by Lewis Prickard (or Piccard) about whom little is known but who may have designed Awe as well. It was financed by Lord Breadalbane and by the trustees of the Forfeited Estates, and cost £650. It is 180 feet overall and slopes from south to north with three segmental spans (48, 42, 30 feet) of decreasing height. They are 22 feet, 18 feet and 15 feet above the water with low piers. The parapet is topped with boulders. It is a peaceful place with the still water of the pool overhung by beeches. To the south east is a three-span embankment taking the road over marshy ground for some 69 yards. It is the same date as the bridge, but only the central arch is more or less as it was. Close by is the unusual and interesting octagonal church of Dalmally designed by James Elliot in 1808.

Proceeding west you will see the seven-span lattice girder railway bridge, built in 1880 for the Callander and Oban Railway, and, near it, Kilchurn Castle.

The BRIDGE OF AWE (*NN 031 298*) over the River Awe is now by-passed by the A85. Although left on one side as long ago as 1938, the bridge and its approach roads are in fair repair. It is a three-span rubble bridge built in 1779 of dark rubble. It stands at an obvious place for a river crossing where the narrow Pass of Brander widens into more level ground and the river is calmer. The arches, from east to west, measure approximately 45, 50, 47 feet. They are footed on low plinth-like piers from which the rise is 21 feet. The parapet is 27 feet above the water, and it splays out into wide refuges for waiting traffic. It was built by the Commissioners of Supply because Lorn, Morvern and the islands were cut off from the Dalmally to Stirling road which was the route to the important Crieff cattle fair. It cost £600 and this was met by the local landowners (£400) plus £200 from the funds of the Forfeited Estates Commission. While being built, with the centering in place, a flood in 1778 swept away the middle arch.

Beyond here is Taynuilt and the interesting Bonawe Furnace. If you park there you can take a pleasant fifteen minute walk to BONAWE BRIDGE (*NN 017 315*). Take the broad path uphill from the Furnace gate and at the cottages above go left and follow the road, which bears to the right. Just before it rises over a small railway bridge there is a metal swing gate on the left. A path goes down the field from here to the bridge which is screened by trees. The man who directed me called it a swing bridge, so I was not expecting what I found. It is a suspension bridge without a fixed deck; so perhaps he should have called it a sway bridge. The old Scots term for such bridges is shaking bridge. This one was built during the construction of the power station for the use of the workmen. It spans the River Awe.

118   Bonawe 'sway bridge' over River Awe

The cables hang from metal pylons in the shape of a tall, wide V with a horizontal bar at the top. The suspender wires, bolted on by links, continue below the deck to a second pair of cables underneath. From the waterside below you will see thåt these cables, anchored in the abutments, curve in the reverse arch of the suspension cables above. Metal girders carry the wooden deck which has an overall width of 2 feet but a walking width of only 17 inches. The sides are made of strong iron mesh attached to a rail. The sides are 4 feet high but they slope outwards so that one walks through a V-shaped net. On the east side is a staircase down, and on the west a gangway entry. This simple, strong bridge is pleasing to look at because of the way it uses the same shape, of a flat bottomed V, in several ways, and because of its unpretentious serviceable materials.

Return to Dalmally and take the A819 to Inveraray.

119   Aray Bridge at Inveraray

If you want to see the romantic taste of the moneyed people in the eighteenth century, you cannot do better than visit Inveraray, where the Duke built his palace-like castle and beautified the surrounding landscape.

The first bridge to catch your eye from the town frontage was designed to do just that. ARAY BRIDGE (*NN 098 091*) spans the Aray where it enters Loch Fyne and makes a focus for the surrounding views. It is most charming on a fine day when the three-fold reflections of two shallow arches and the circular oculus gleam below the wooded hillside.

The best place to see it is from the riverside where it strikes one as uncommonly handsome. The voussoirs and the parapet are a pinkish stone. The spandrels have at some time been coloured a soft green which is wearing off. The large oculus right through the central spandrel is circled by a stone rope. The arches have square archivolts.

However, when I am actually on the bridge, I find it to some extent a let down because the parapet detailing is clumsy. What reads as a smooth curve is seen to be made of seven straight runs. The sections of balustrading (apparently inserted on the express wish of the Duke) do not go with the rise of the road, and, from the inside, are rather awkward and ugly. Of course, this did not matter when seen from afar as part of a romantic landscape.

From the castle the arches frame loch and hillsides, and, in a boat, they frame sections of the castle grounds. The town, away to the right, cannot be seen by those for whom the bridge was designed. It was built in 1776 by Robert Mylne, and is sometimes called New Bridge.

120   Garron Bridge

From the bridge you have splendid views of the castle, and of the town which is a handsome long, low rectangle of white buildings, diversified with arcading and the brown bell tower, and finished off with the dark wooden legs and white railings of the jetty.

GARRON BRIDGE (*NN 114 101*) at the head of Loch Shira over the outlet of the Gearr Abhain was designed by Roger Morris and built in 1748, so it is the oldest of the ornamental Inveraray bridges. It too was meant to be a focal point in a view, but, being so far from the town and castle and somewhat dwarfed by the hillsides above it and water below, is less successfully placed than Aray. However, close to, it is a pretty bridge especially when reflected in the green and silver water. The beach shelves steeply under it. The arch springs from the high tide level. It is a high humped bridge with semi-hexagonal abutments which extend up into square refuges and are decorated with entablatures, and ball and

121   Dubh Loch Bridge, with passer-by watching the swans

finial ornament. The central section of the parapet matches the
abutment tops and each side there is a section of balustrading. The
string course is denticulate. An interesting detail is the way the spandrel
courses run—not horizontally as usual but diagonally raying out from
the archivolt. If you walk over the bridge you will notice the
undecorated backs of the entablatures and a generally much plainer air.
It was not important that the inside be decorative because it was not
part of the arranged view.

Very near here is another bridge. Where the road swings round the
head of Loch Shira there is a tarmac entry to an estate road. You can
park here and walk the few yards down to **DUBH LOCH BRIDGE** (*NN 114
105*), sometimes called Sword Ford Bridge. It spans the Garron. It also
was designed by Robert Mylne and built in 1757, the contractor being
a man called Tavish. Sitting over the Garron where it leaves Dubh
Loch, its strong, severe lines are in keeping with the scenery of hill and
loch. It is a single, segmental arch with double voussoirs which widen
at each end. The parapet is straight with cruciform slits. The abutments
rise into towers with a single large castellation each. Linking spandrels
to abutments, there are quarter circle arcs, like sections of a tower, also
with a cut in their parapet. These details give it a vaguely military air.
The towers have two rows of corbelling and, at their feet, a splayed skirt
of sloping stone. There is a denticulate string course. At some time the
bridge has been harled and this remains on the voussoirs which are
therefore hard to see in detail. The wing walls are long, sloping and
curved, and finish in square towers. The stone is dark.

122   Garden Bridge, or Frew's Bridge, in the Castle grounds

Altogether, it fits one's ideas of a bridge suitable for a castle in the mountains. When I was there in October, fishermen's boats were drifting across the loch and autumnal sunlight showed the green and gold tapestry of leaf and bracken clothing the nearer hillsides. A woman strolling there in the sun looked through the parapet nick at a family of swans floating by. It seemed remote from the elegance of Aray and Garron, although so near.

GARDEN BRIDGE (*NN 095 095*) in the Castle grounds is the most charming of the classical bridges connected with Inveraray. It has some of the finest masonry work that I have seen; so it seems fitting that locally it is called Frew's Bridge after the mason David Frew. It was designed by John Adam in 1761, and stands over the Aray.

It is built throughout of a greenish grey stone, and each separate block is scored with fine diagonal lines, like the Castle itself. This is called broaching by Scots masons. It is almost unnoticeable, and yet it gives a surface texture with innumerable fine shading lines.

There is a single elliptical arch of 60 feet and a balustraded parapet. The abutments have semi-circular towers which at the road level form refuges. These have stone seats inside. The voussoirs are in a double row with an archivolt. The spandrel courses ray out as they do at Garron. The string course is denticulate and juts out to look like a shelf from above. In both spandrels are blind oculi edged with moulded stone. The courses inside these oculi are horizontal, not radiating. Where the spandrels and voussoirs come up against the abutments there is a

123   Garden Bridge, detail of refuge and seat

fascinating juxtaposition of differing stone work (see illustration 000). The amount of detail and ornament on this bridge is large and yet, in no way, obtrusive or bothersome. It is, rather, completely harmonious.

On the river bank you see all this beautiful workmanship to best advantage. Down there you will notice that the sandstone wing walls have pillared classical doorways of fine stone which give access to passageways under the bridge linking the water meadows.

On the bridge, to which there is a steep rise, you will see that the problem of setting a flat top on a curve has been better managed than at Aray Bridge. Here at the middle there is one course of stones between balustrade and road which gradually increases to five courses at the end of the bridge refuge. There has been a gate across at one time and the four-fold wooden gateposts with iron finials remain. From here there is a charming view up to the big round white dovecot.

It is interesting to note that at this date of 1761 no other elliptical arch bridge had been built in Britain, although John Adams was building another at Dumfries House in Ayrshire and Mylne had already designed, but not yet built, Blackfriars Bridge over the Thames.

If you follow the estate road past the dovecot you will come to FOAL'S BRIDGE (*NN 087 109*) over the Aray. Its curious name is probably a corruption of a Gaelic word meaning ravine, for there is one just above it. The designer may have been Mylne, striking a rustic note, but the mason was certainly a William Douglas, who gave an estimate for the work of £60. 3*s*. 0*d*. Having started it in 1751, he left it half done to work

124   Garden Bridge, masonry details

on the Great Inn at Inveraray and it was not finished until 1757. It is a simple bridge of rubble with a span of 40 feet, hardly coursed but with rounded coping stones. There is a noticeable slant uphill to the roadway. The river sides are much overhung by beeches, oaks, alders and rhododendrons.

In complete contrast to these five eighteenth-century bridges erected to beautify a rich man's estate, are some little bridges in Glen Aray just off the A819 to the east, most connected with the old military road built between 1750 and 1757 under Major Caulfeild. This road ran on the left bank of the Aray out of Inverary.

To find them look out for the Three Bridges Pottery and, about half a mile north, there is a turning right which is closed off. You can park in this entry and then walk straight ahead, climb the deer fence by the stile and you will see two bridges (*NN 090 130*). One is a ruinous wooden bridge perched on the rocks over a waterfall. Next to it is the least rigid, though perfectly safe, bridge that I have found. It was one of only two that my spaniel refused to cross, preferring to swim. It was built in memory of Ian Birkmyre, a man who came here for many years to fish and loved the place. The construction is of the simplest. Cement abutments anchor metal poles from which is slung a wire net cradle which is given a flat bottom by means of three squared U-shaped sections bolted on and some wood slats. When I crossed it, the day was calm and the river low. I am told that, in times of storm, the spray splashes over one's head.

125   Birkmyre's suspended cradle

From Mr Birkmyre's bridge, follow the forestry track to the right downhill until you come to a gate in the deer fence. Beyond, in the field, you can see a small grass covered stone arch bridge (*NN 092 126*) which is on the line of the military road and may be an original bridge repaired by estate workers.

Back at the car, drive on northwards for just under a mile where, near the track on the left to a house called Stronmagchan, the old road forms a layby. Park here and look, on the opposite side of the road, for an iron gate in the trees. This leads down what is clearly an old road with tall trees either side and up to a knoll on which stands a ruined building which was once a school. Below is another small stone arch bridge, also grassed (*NN 087 140*) and probably part of the military road. Faced with the solitude and woodland peace of this forgotten school and road, it is hard to summon up pictures of the soldiers building it and the grateful travellers using it, as did Boswell and Dr Johnson going south from Mull in 1773. In fact for this stretch the parties of soldiers were augmented by tenants of the various parishes. This work was statute labour. All tenants had to give six days' work a year on highways; but much argument ensued between the army and the County authorities who said that such labour was intended for county and not military roads. Often there was no clear distinction between the two, the military road being the only road, and gradually responsibility for it was assumed by the civil authorities. This partly explains why so many old roads are described as military roads when they do not figure in Wade or Caulfeild's maps. At a later date soldiers may well have been employed repairing them. This may be the cause of the dispute over the Contin to Poolewe road discussed in Chapter 9.

There are two other country bridges worth finding to the south west of Inveraray, both crossing the Douglas Water. Take the A83 towards Lochgilphead and stop, after about 3 miles, at the Argyll Caravan Park. Leave the car here and take the very rough track to the right which leads downhill to DOUGLAS BRIDGE (*NN 072 091*), a pleasant hump-backed stone arch dated approximately 1785. It is rubble, hardly coursed at all, with narrow voussoirs and a square-ended parapet with rounded coping stones. The track you came down is the old main road from Inveraray to Furnace which ran beside the loch and can still be walked. The right hand turning here will lead you to another old bridge even more prettily sited but it can also be reached by car. Continue south about another mile and stop on the left beside an entry to a lane gated with an orange metal gate. (Across the road are the footings of the old Bridge of Douglas.) Follow the track past some ruined cottages and then take a field path left in front of an abandoned farmhouse. In another 50 yards you will find the bridge, a rubble segmental arch over a low but pretty waterfall overhung by beeches (*NN 058 048*). This is known locally as the Roman Bridge, which is even more startling to hear than the ubiquitous Wade bridge. It is almost certainly a late-eighteenth-century bridge taking the cross track which we have already seen. In the autumn with the overnight mist lingering low on the woodland paths and the strengthening sun striking gold and scarlet from the beech leaves, both these bridges and the path between them were pleasant places to linger.

From Inveraray the A83 runs east and the bridges on it are at the outer edge of the area covered by this book. Two are of interest.

FYNE BRIDGE (NN 193 127) over River Fyne at the head of Loch Fyne is a splendid by-passed structure on the line of the military road and built about 1745. It has an elegance these older bridges rarely achieve. There are four low segmental spans outlined by dressed stone voussoirs and an archivolt, although the rest of the bridge is coursed rubble, excepting the cutwaters and the pier footings. The piers extend up into rectangular towers which make shallow refuges. The main spans are approximately 33 feet but the arch nearest the left bank is lower than the other three. The abutments are not symmetrical but both join the spandrels at an unusually sharp angle and with a corbelled stone set crosswise. The parapet has a flat top, with some coping stones being concrete replacements, but the old ones are liberally inscribed with initials and dates, some quite old. The bridge crosses a pebbled, shallow part of the river just upstream of an undistinguished modern bridge.

126   Fyne Bridge, note the dressed stone voussoirs

Behind rise the sloping hillsides cut by the sharply shadowed little runnels of numerous burns that only flow in rainy weather.

Fyne Bridge is handsome and unusual, its good looks chiefly due to the voussoirs being highlighted and the square refuges.

In Glen Kinglas is **BUTTERBRIDGE** (*NN 234 095*) built in 1745, on the military road Caulfeild made, by a Dunkeld mason called Thomas Clark. It is by-passed and lies in a hollow and can easily be overlooked. It is a sturdy little humped bridge, its roadway now tufted with grasses, and quite in keeping with the rocky, wild mountain pass. The segmental arch has a span of 30 feet approximately. The voussoirs are uneven and many of the top ones are almost vertical suggesting that mortar was not used in the building (see page 122). The abutments are set on the river bed. The old road runs away, nothing more now than a rushy track. It is set under the steep-sided peak of Beinn an Lochain and overlooked also by Beinn Luibhean, Beinn Ime and Beinn Chorranach. When I was there in the evening the carpark was full of weary climbers and walkers thankfully regaining their vehicles. The sun, already below the hilltop, threw dramatic shadows on the worn green velvet hillsides.

The last section of this chapter deals with the Crinan Canal and the bridges in that area which is as far south as this survey goes. The boss of Bute can be approached either on the speedy A83 down the coast of Loch Fyne, or via the wooded B840 from Cladich to Ford.

At Bridgend is the by-passed **KILMICHAEL BRIDGE** over the River Add

(*NR 853 926*). It is a handsome rubble bridge with two main segmental spans and two flood arches. It was built in 1737 at the Shire's expense.

Near here a side road, the B8025, leads directly to Islandadd and the canal. Some way down is POLTALLOCH EAST LODGE BRIDGE (*NR 825 960*), an ornamental sandstone bridge at one of the entrances to the, now ruined, but still handsome Poltalloch House. It has a single elliptical span, a decorative, diamond lattice and stone balustrades that curve away at each end. The abutments are finished with towers topped by pinnacles. Although the spandrels are rubble, the rest is finely dressed stone. There is also a decaying but very ornate wooden gate. The date is 1855.

CRINAN CANAL BRIDGES	1794–1809;	John Rennie	*NR 793 938*
	and 1817;	Thomas Telford	and
	and later		*820 912*

The canal was the first built in Scotland and made to cut out the difficult passage round the Mull of Kintyre and so aid trade between Glasgow and the islands. Rennie's salary was set at 200 guineas a year but he halved it on the understanding that he could not spend as much as a month a year in Argyll. The wages of the labourers in 1798 were sixteen shillings a week for masons and twelve for labourers. The original capital raised was £92,000 but it cost more. It proved difficult to construct partly because of the unstable nature of the ground in the western section, the Great Moss between Bellanoch and Crinan, and also because Rennie found the work hard to control at a distance, so that some poor and often desultory work was done. It was left to Telford in 1817 to do some thorough repairs. He re-puddled the leaking walls and replaced lock gates and built a longer breakwater at Ardrishaig. His deputy was John Gibb for this work. Those interested can read a full account in A D Cameron's booklet. The canal is now used not by trading or passenger ships but by yachts. Two thousand used it in 1987.

There are six swing bridges. The two most interesting are near Crinan (*NR 793 938*) and at Dunardy over lock 11 (*NR 820 912*).

DUNARDY TRAVERSING BRIDGE probably deserves the over-worked adjective unique. At first there was a swing bridge here but it was found to be too heavy for the lock's rather shaky foundations, so this lighter bridge was installed in 1900. It is, in effect, a single cantilever supported by steel rods connected to a pair of lattice pyramid pylons. The deck is wooden. The bridge keeper winds it backwards on rails, more like a flat drawbridge than anything else. There is a small wooden platform for the keeper to stand on as he turns the wheel and he is himself moved

127 Dunardy Bridge over the Crinan Canal. Note platform with wheel

128 Crinan Swing Bridge being opened

129   Islandadd Bridge. Note the wrought iron truss below

backwards with the bridge. The diminutive size makes it seem more of
a toy than a real working bridge, but quite a fair amount of local traffic
uses it.

CRINAN SWING BRIDGE, built in 1870, is the oldest remaining one with
the original mechanism. There is a large, very oily, hand-operated
ratchet wheel, and under a trap door in the planking a greasing point.
Before the keeper can wind the handle, he has to pull a tall lever which
lifts the bridge off its 'wedges'. It is a pretty construction with wrought
iron railing in a diagonal cross pattern. The abutments are stone curved
walls painted in black and yellow stripes. Across the bridge you reach
the towpath which goes on to the Crinan basin or back over the estuary
water to Bellanoch. The cottage here was known as Puddler's Cottage
as the man, who was expert at this job of stopping leaks with clayey
mud, lived here.

At Bellanoch there is a similar bridge that has been modernised and
beside it over the Add is Islandadd Bridge.

ISLANDADD BRIDGE (*NR 805 925*), at the junction of the B841 and
B8025, was built in 1851 by John Gardner. It is one of the few cast iron,
parallel girder bridges in our area, and the largest by some way. It has
five spans supported by narrow stone piers and railed with an elegant
and delicate cast iron balustrade. Close to, you will see that at some time
it has been underpinned with wrought iron truss girders. These are quite
decayed in places with rust. The abutments, square-topped like the
piers, are solidly built of rustic ashlar stone, with battered walls,

rounded steps and big skirts of sloping stone at water level. The tide rises and falls, covering and uncovering these and the small triangular cutwaters.

I find this an especially pleasing bridge, its long, low line well adapted to the wide waters. The estuary, backed by the Moine Mhor through which the River Add drains to the sea, makes a beautiful end to the wooded and craggy countryside of this peninsula. In early morning and evening, the light and the reflections in the water of the rushes, trees and hillsides are lovely. One of the happiest short walks is westwards along the towpath from Islandadd Bridge to Crinan. It strikes out across the calm water on a long, grassy causeway which offers serenity and freedom.

Glossary

abutments	the side supports of a bridge
archivolt	moulded stone edging to an arch ring: only seen in bridges with some ornament
ashlar	dressed stone with squared edges: there are various styles and various names for them
balustrade	a row of low pillars forming a parapet
batter	the slope (of a wall) from the perpendicular
cantilever	a projecting arm fixed at one side only and not requiring support below (see page 000)
cartouche	a decorative panel, usually over the keystone of an arch, and often with an inscription
centering	a wooden frame to support a stone arch while being built. Also called a falsework (see Figure 5)
clam	a single span bridge formed of flag stones
clapper	a clam bridge with several spans and low, drystone piers: often called a flag bridge in Scotland
coping	the top of a parapet, often projecting or decorated
corbel	a projection from a wall to support a weight above
cutwater	additional footing added to a pier at water level and into the current, to divert the force of the water's flow: often triangular in shape, and extending upwards some distance
dentils	a row of projecting, single, square stones, usually below a string course, and looking somewhat like teeth: so, denticulate
drystone	masonry built without mortar
elliptical	oval (see Figure 4)
extrados	the upper curve of an arch
girder	a large, longitudinal beam, frequently of metal
impost	the upper stone course of a pier or abutment below the springing line, sometimes decorated with a moulded rim.
intrados	the inner curve of an arch
keystone	the central stone of a voussoir ring, sometimes larger than the rest
oculus	a circular or oval ornament (like a wreath or an eye) in the spandrel, sometimes hollow, always with a moulded rim
overbridge	used of a bridge which passes over a railway line
parapet	the low wall, or railing, which edges the roadway of a bridge
pier	column supporting the central section of a bridge: when built in water a pier has considerable foundations below the river bed
pozzulan	type of cement used by the Romans who used volcanic dust from Pozzuoli

213

precast   (of concrete) shaped in moulds before being built into the structure
pre-stressed   (of concrete) given tensile strength before exposure to loads by means
                     of tightening inner steel bars
refuge   alcove in the parapet for the safety of pedestrians, usually formed by a
                     half-pillar being built above a pier or abutment
reinforced   (of concrete) with steel stiffening bars incorporated
revetted   (of river banks) faced with stone to strengthen them, and to channel the
                     water more effectively
segmental   the curve of only part of a semicircle (see Figure 4)
soffit   the voussoir courses continued under the bridge; the under surface of an
                     arch; the vault
span   the distance between two abutments, or between abutment and pier
spandrel   the courses of masonry over the arch ring, or rings, which extend up to
                     the parapet and out to the abutments
springing   the lowest point of the arch, where the curve begins on the abutment or
                     pier
starling   an underwater guard made of piles and stones round the cutwater to
                     form an extra protection for the pier against floating objects
truss   a beam constructed of short lengths of wood or metal, usually arranged
                     in triangular or diamond shapes (see Figure 8)
turnbuckle   metal sleeve with screw threads at each end running in opposite
                     directions, to allow a cable to be tightened
underbridge   used of railway bridges when the rail is carried on the bridge over the
                     obstacle
vault   the roof curve of an arch
voussoirs   the half-circle of stones forming the outer edge of the vault: the arch ring
wingwall   a continuation of the abutment side walls, often providing an embank-
                     ment for the approach road

List of Bridges

New Spey, Grantown
Newtonmore
Tongue

Clapper Bridges
Aultbea
Badachro
Killin
Riconich
Torridon

Bridges plus walks (mileage there and back)
Athnamulloch (12)
Bonawe (1½)
Eye of the Window (4)
Glen Einich (8)
Glen Muick (1)
Glen Roy (5)
Moy Swing Bridge (2½)
Old Forss (1)
Sandwick (1½)
Sluggan (1½)
Torridon House Bridge (2)

25 Bridges not to be Missed
Athnamulloch

Altnaslanach
Ballindalloch
Broomhill
Cambus O' May
Craigellachie
Craigmin
Fleet
Gairnshiel
Garden Bridge, Inveraray
Garvamore
General's Wells Bridge
Glen Einich
Glenfinnan Viaduct
Flen Livet Packhorse bridge
Glen Loy Aqueduct
Invercauld
Kessock
Kylesku
Little Garve
Newtonmoe
Plodda
Sluggan
Speybay Viaduct
Torridon clapper
Whitebridge

If you can only see one—make it Craigellachie

130   A bridge no more: the ultimate in bridge conversion, near Fort Augustus

Bibliography

Album of Photographs of Inverness–Perth Road Bridges 1924–27 by courtesy of Sir Owen Williams & Partners (Privately owned)

Billington, David, *The Tower and the Bridge* (Basic Books Inc, 1983)

Bracegirdle, B and Miles, P, *Thomas Telford* (David and Charles, 1973)

Buchanan, A and Jones, S K, 'The Balmoral Bridge of I K Brunel', *Industrial Archaeological Revue*, Vol 4, No 3 (Autumn 1980, OUP)

Burgess, R and Kinghorn, R, *Speyside Railways* (AUP, 1988)

Calder, Sinclair, 'Industrial Archaeology of Sutherland' (unpublished PhD thesis lodged with Inverness Public Library)

Cameron, A D, *The Caledonian Canal*, (Melven Press Ltd, 1983)

——, *Getting to Know . . . The Crinan Canal* (1985)

Casson, Hugh, *Bridges* (Chatto and Windus Ltd, 1963)

Close-Brooks, Joanna, *Exploring Scotland's Heritage, The Highlands* (RCAHMS, HMSO 1986)

Curtis, G R, 'Roads and bridges in the Scottish Highlands: the route between Dunkeld and Inverness, 1725–1925', *Proceedings of the Society of Antiquaries of Scotland*, 110 (1978–80)

de Mare, Eric, *The Bridges of Britain* (Batsford, 1954)

Fenton, A and Stell, G (ed), *Loads and Roads in Scotland and Beyond* (John Donald)

Gibb, Alexander, *The Story of Telford* (Alex MacLehose and Co, 1935)

Gordon, J E, *Structures* (Pelican, 1978)

Graham, Angus, 'The Military Road from Braemar to the Spittal of Glenshee', *Proc of Society of Antiquaries of Scotland*, Vol 97.

Haldane, A R B, *New Ways through the Glens* (David and Charles, 1962 and 1973)

Henry, D and Jerome, J, *Modern British Bridges* (CR Books Ltd, 1965)

Hopkins, H J, *A Span of Bridges* (David and Charles, 1970)

Hume, John R, *The Industrial Archaeology of Scotland: II The Highlands and Islands* (Batsford Ltd, 1977)

——, 'Telford's Highland Bridges', Chapter 8 in *Thomas Telford: Engineer* (proceedings of seminar at Coalport Museum, April 1979, 1980)

——, 'Scottish Suspension Bridges', *The Archaeology of Industrial Scotland Forum 8* (EUP, 1977)

——, 'Cast iron and Bridge-building in Scotland', *Industrial Archaeology Review*, Vol 11, No 3 (1978)

Lauder, Thomas Dick, *Account of the Great Floods* (Adam Black, 1830)

Mitchell, Joseph, *Reminiscences of My Life in the Highlands, Vols 1 and 2* (David and Charles, 1971)

Nixon, L A and Robinson, P J, *British Rail North of the Border* (Ian Allan, 1983)

Pride, Glen, *Glossary of Scottish Building* (Scottish Civic Trust, 1975)

RCAHMS, 'Argyll: an Inventory of the Ancient Monuments', *Vol 2 Lorn* (HMSO, 1975)

Reiach Hall Blyth Partnership, 'Footbridges in the Countryside', *Design and Construction* (Countryside Commission for Scotland, 1981)

Ritchie, G and Harman, M, *Exploring Scotland's Heritage, Argyll and the Western Isles* (RCAHMS, HMSO, 1985)

Rolt, L T C, *Thomas Telford* (Longmans 1958, reprinted Pelican 1979)

Ruddock, Ted, *Arch Bridges and their Builders, 1735–1835* (CUP, 1979)

Salmond, J B, *Wade in Scotlnad* (Moray Press, 1934)

Scott, Alistair, *Bridges in Moray* (Moray Field Club, 1981)

Shepherd, Ian A G, *Exploring Scotland's Heritage, Grampian* (RCAHMS, HMSO, 1986)

Smout, T C, *History of the Scottish People, 1560–1830* (Fontana, 1969)

Southey, Robert, *Journal of a Tour in Scotland in 1819* (Murray 1929, facsimile Thin 1972)

Stamp, Gavin (ed), *Sir Owen Williams, 1890–1969* (Architectural Association)

Steinman, David and Watson, Sara, *Bridges and their Builders* (Dover Pub Inc and Constable, 1957)

Taylor, William, *The Military Roads of Scotland* (David and Charles, 1976)

Telford, Thomas, *Atlas of his Works*

Thomas, John, *Forgotten Railways of Scotland* (David and Charles, 1976)

Index

(The page numbers given refer to the main entries.)

219

THE SUN RISING THROUGH VAPOUR
TURNER'S EARLY SEASCAPES

Paul Spencer-Longhurst

The Barber Institute of Fine Arts
The University of Birmingham

24 October 2003–25 January 2004

Third Millennium Publishing
in association with
The Barber Institute of Fine Arts

First published in 2003 by Third Millennium Publishing,
an imprint of Third Millennium Information Limited
First Floor
2 Jubilee Place
London
SW3 3TQ, UK
www.tmiltd.com

ISBN: 1 9803942 25 X

All measurements are in centimetres
Where applicable, height precedes width precedes depth

Edited by Catherine Walston
Copy-edited by Honeychurch Associates, Cambridge, UK
Designed by Third Millennium Information Limited, London
Produced by Third Millennium Publishing, an imprint of
Third Millennium Information Limited

Reprographics by News S.p.a., Italy

Printed and bound in Italy by Sfera International

Front Cover:
The Sun rising through Vapour, detail, c. 1809 (cat. 7)
Signed lower right: *J.M.W. Turner, R.A.*
Oil on canvas, 69 x 102 cm
The Barber Institute of Fine Arts, The University of Birmingham

Back Cover:
Fishermen upon a Lee Shore in Squally Weather, RA 1802 (cat. 2)
Oil on canvas, 91.5 x 122 cm
Southampton City Art Gallery

Frontispiece:
William Holl the Younger (1807–71) after Turner,
Self-Portrait, published 1859–61 (cat. 25)
Engraving, 16.4 x 13 cm
Tate Britain

CONTENTS

PREFACE

This exhibition has two aims. Firstly, it explores Turner's initial fifteen years or so as a painter of the sea. Secondly, it seeks to set his work in the context of marine painting in England at a pivotal moment in its development. The period from about 1795 saw Turner enthusiastically experimenting with depictions of the sea as a major expressive vehicle in both oil and watercolour. Simultaneously he developed a fascination with the elements of sky and sea and their inhabitants, which continued to grow throughout his life. The group of sea-pieces upon which he embarked early in his career is crucial to an understanding of his oeuvre as a whole. His first exhibited oil painting, *Fishermen at Sea* of 1796, was described at the time as masterly, and his early reputation was founded on a series of dramatic sea-pieces that he regularly showed at the Royal Academy, the British Institution and in his own gallery until about 1810.

It was water, more than any other subject, that allowed Turner's imagination freedom to innovate and to transcend mere naturalism. It was water that boosted his self-confidence through familiarity, from his earliest years on the banks of the Thames and at Margate. His ability to depict the endless motion and power of the sea often acts as a foil to the frailty of mankind and its vessels. Indeed, the fragility of man's hold on the world is a theme that Turner continued to explore throughout his career. However, rather than emphasize once again the stormy aspect of Turner's marines and his large set-pieces, the present exhibition pays special attention to more serene and smaller seascapes, where the ocean is depicted as a source of food, defence and prosperity. It focuses particularly on *The Sun rising through Vapour*, an early marine by Turner which was acquired by the Barber Institute in 1938. This picture, like its more famous cousin of the same title in the National Gallery, reveals a very different side of the painter's imagination. Quiet and reflective, both confirm Turner as the successor to such masters of ineffable calm as Claude, Cuyp, Van de Cappelle, Willem van de Velde the Younger and Vernet.

A natural hiatus occurred with Turner's completion of *The Wreck of a Transport Ship* (Calouste Gulbenkian Foundation, Lisbon) in about 1810 and the years beyond fall outside the scope of this exhibition. Over the two decades that followed, Turner ceased to produce marines regularly, with a few notable exceptions such as *Dort or Dordrecht* (1818, Yale Center for British Art). Only in the 1830s and 1840s did he take up the genre seriously once more. For these reasons it is especially appropriate that this exhibition should run concurrently with *Turner's Britain* at Birmingham Museum and Art Gallery and *Turner: The Late Seascapes* at Manchester Art Gallery. Despite his pre-eminence, Turner is a painter severely under-represented in public collections outside London. It is our fervent hope that this exhibition will help to right that injustice.

Richard Verdi
DIRECTOR

Fig. 1
Fishermen at Sea ('The Cholmeley Sea-piece'), RA 1796, Tate Britain

ACKNOWLEDGEMENTS

Among the many contributors to the preparation of this exhibition and catalogue I would like to acknowledge with particular appreciation the co-operation, assistance and enthusiasm of Professor Brian Allen, John van Boolen, Dr David Blayney Brown, Dr Susan Foister, Dan Giles, Dr James Hamilton, Yvonne Locke, Susan Morris, Rosemary Poynton, Dr Geoff Quilley, Eric Shanes, Tessa Sidey, Professor Richard Verdi, Dr Ian Warrell, Commander Alastair Wilson, RN, and Sophie Wilson. I am deeply grateful to the Trustees of the Barber Institute for their moral and financial backing, and I very much appreciate the generous forbearance of my colleagues at the Barber Institute over my frequent absences during the past year. The exhibition could not have been brought to fruition without the constant encouragement and support of Sue Spencer-Longhurst and Rose and Flora.

To all the lenders listed opposite the Barber Institute is deeply indebted. There could be no major exhibition on Turner without the willing co-operation of Tate Britain, nor one involving marine painting without that of the National Maritime Museum, Greenwich. We also recognise that the precise nature of the exhibition makes us highly dependent upon the goodwill of individual lenders, for whom the temporary loss of one work may be a major sacrifice.

Paul Spencer-Longhurst

ABBREVIATIONS

The following abbreviations are used throughout the catalogue:

BI	Exhibited at the British Institution
Exh. cat.	Exhibition catalogue
RA	Exhibited at the Royal Academy
TB	Turner Bequest, Tate Britain
TG	Exhibited in Turner's Gallery

LIST OF LENDERS

Birmingham Museums and Art Gallery	21
British Museum, London	23, 24
National Gallery, London	6
National Maritime Museum, Greenwich	16, 17, 18, 19, 22
Private Collection	15
Richard Green, London	8
Southampton City Art Gallery	2
Tate, London	1, 4, 5, 9, 10, 11, 12, 13, 14, 20, 25, 26
Yale Center for British Art, Paul Mellon Collection	3

TURNER'S EARLY SEASCAPES

It is easy to forget that during his lifetime Turner was regarded as 'the great sea-painter'. The later criticism of John Ruskin substituted landscape as the focus of his reputation, but this change was by no means universal at the time of the artist's death in 1851. The following year a telling comment was made by his first biographer, John Burnet, in a passage that has since become famous:

'Of all the variety of subjects which the versatility of Turner's genius led him to paint, there were none which he seemed to be so completely master of, or execute with greater care or more spirit, than his sea-pieces, especially when the tempest-tossed waves threaten to "swallow navigation up"; nothing can exceed the appearance of turbulent motion with which he imbues them; their forms can only be caught sight of ere they hurry into confusion and become lost.'[1]

Burnet went on to name what he called *The Wreck of the 'Minotaur'* (fig. 2) and certainly had in mind *The Shipwreck*, the greatest of Turner's tempestuous marines that made his name in the first decade of the nineteenth century (fig. 6). A print of the latter was reproduced as the frontispiece of Burnet's biography. Nor was he alone in his judgement, which had been anticipated by Charles Robert Leslie, the painter and biographer of Turner's rival, Constable. Leslie's son, Robert, himself a painter of marines, told Ruskin how, when staying at Petworth as the guest of Lord Egremont in about 1832, his father came upon Turner and named him as 'the great sea painter'. Writing in 1884, Ruskin commented, 'I have

Fig. 2
The Wreck of a Transport Ship, 1810–12
Calouste Gulbenkian Foundation, Lisbon

put "sea" in italics, because it is a new idea to me that at this time Turner's fame rested on his marine paintings.'[2] Yet Ruskin, Turner's greatest apologist and interpreter, believed that his protégé had no equal in painting rivers or seas, storm or calm, surface or depth, among either contemporaries or predecessors.

No other artist could match Turner as a painter of the sea in all its moods, although it was its tempestuous side that struck the deepest chord in him; for storm and shipwreck appealed to the strain of pessimistic sensationalism in his nature. Turner's 'first oil picture of any size or consequence was a view of flustered and scurrying fishing-boats in a gale of wind off the Needles,

Fig. 3
Dunstanborough Castle: Sunrise after a Squally Night, RA 1798
National Gallery of Victoria, Melbourne, Australia

Fig. 4
Boats carrying out Anchors and Cables to Dutch Men-of-War in 1665, RA 1804
Corcoran Gallery of Art, Washington, D.C.

which General Stewart bought for £10.'³ His earliest commissioned oil of any importance was the missing 'Mildmay Sea-piece', painted in 1797 for Sir Henry St John-Mildmay, Bart; but for several years from 1796 his exhibited oil paintings were all situated on the coast, or in close proximity to extensive areas of water. These include topographical works (fig. 3), fishing scenes, views of seaports, shipwrecks, real or imaginary incidents from the seventeenth-century Anglo-Dutch wars (fig. 4), and the biblical *Deluge* (fig. 5). As his underlying theme, the sea was to become the barometer of Turner's changes in style and the *locus* of his evolving Romantic spirit. Almost a quarter of his catalogued oil paintings have the sea as their predominant subject, and this number increases significantly if classical and harbour scenes and views of Venice are included.⁴ The majority of these sea pictures are concentrated in the two periods at the beginning and

Fig. 5
The Deluge, TG 1805 (?), Tate Britain

Fig. 6
The Shipwreck, 1805, Tate Britain

end of his career and the present exhibition is concerned with those leading up to 1810. The first two decades of Turner's productive life witnessed the execution and display of some of the greatest and most monumental of his marine pictures, culminating in *The Wreck of a Transport Ship* (fig. 2). Only in his last two decades would his output of seascapes rival this early series.

Apart from Turner's personal fascination with the sea and rivers, which took its origin from the circumstances of his childhood, there were general reasons for the increasing popularity of seascapes at this time. The prolonged period of naval warfare, which had begun during the American War of Independence and ended with Napoleon's defeat at Waterloo, led to an increasing demand for marine painting, especially pictures of famous ships and battles at sea. The strength of the British Navy was vital to the defeat of France and the English Channel presented the only barrier against French military ambitions. But isolation from Europe

led to a siege mentality that nurtured a national concern for patriotism, security and prosperity. The fishing industry was indispensable to Britain during these years as a source of both food and employment. A large proportion of Britain's population lived in seaports or in coastal villages and the sea provided direct or indirect employment for many people. The period also witnessed a change in public attitudes towards the coast, which was traditionally regarded as the wild preserve of fishermen and smugglers. Visits for sea-bathing became a popular middle-class pastime, especially at Brighton, which developed rapidly from a small fishing village into a fashionable resort, in the wake of the Prince Regent's presence from 1783. Towns such as Dover and Margate quickly followed suit. Picturesque subject-matter also began to be sought among the fishing boats, beaches and harbours, by visiting painters and watercolourists such as Joshua Cristall, who spent time at Hastings in 1807.

Public fascination with the sea was largely driven by the dangers it contained and the necessity for risking exposure to these, in an age when much long-distance travel involved a sea-voyage. The dread of shipwreck and drowning was little short of obsessive, and with reason, when about 5,000 British people were drowned at sea each year.[5] One of the best-known disasters of the period was the sinking of the *Abergavenny* off Weymouth in 1805, with the loss of its captain, the young John Wordsworth, brother of the poet. Public awareness was increased by newspaper accounts, the theatre, commemorative prints and sometimes by paintings.[6] James Northcote's *Wreck of the 'Centaur'*, for example, was exhibited at the Royal Academy in 1784. Literature too played an important part. Daniel Defoe's *Robinson Crusoe*, first published in 1719, was the *fons et origo* of

Fig. 7
Dutch Boats in a Gale ('The Bridgewater Sea-piece'), RA 1801
Private Collection on loan to National Gallery, London

shipwreck tales. Throughout the eighteenth century it had gone through many editions and it is almost certain that Turner would have read it by his late teens. With its narrative range over virtually every possible maritime experience, it linked the dangers of the deep with the presence of God through the overwhelming power of nature. It gave rise to a whole genre of shipwreck tales which proliferated in the early nineteenth century, either in the form of pamphlets or of multi-volume anthologies, such as the Reverend James Stanier Clarke's two-volume *Naufragia*, published in 1805.[7] The sea and its dangers were also a preoccupation for early Romantic poetry, from William Falconer's *The Shipwreck*, first published in 1762, to Coleridge's *Ancient Mariner* of 1798, and later Byron's *Don Juan*.[8] As early as 1797 *The Shipwreck* was mentioned in connection with Turner's 'Mildmay Sea-piece'.[9] Turner, who was 'fond of talking poetry', read

these works with critical attention and was well versed in Milton, Thomson, Mallet and others.[10] His own attempts at poetry also originated in the first decade of the century, and the incomplete versifications known as *The Fallacies of Hope* are full of references to natural disasters and mankind's helplessness in the face of nature.

Joseph Mallord William Turner was born on 23 April 1775, the son of a barber and wig-maker of Maiden Lane, Covent Garden, London. He was encouraged to develop an interest in art by his father, who displayed his early drawings for sale. Turner's training was, however, far from orthodox. It included flower painting, rudimentary perspective, adding skies to architectural drawings, copying prints and drawings and putting washes within topographical outlines. His association with water and the sea began at an early age. Born within 300 metres of the River Thames, its varying moods by day and night never ceased to fascinate him (see cat. 1). In 1786, during the fatal illness of a younger sister, he went to live with a maternal uncle at Brentford, becoming familiar with the reaches of the river west of London. Soon afterwards he stayed with relatives of his mother in the busy port of Margate on the Kent coast. He spent some time that year as a pupil at a school in Love Lane, and it was at Margate that he 'first learnt the physiognomy of the waves.'[11] Back in London, at the age of fourteen, he started to attend the Royal Academy Schools and to study figure-drawing, first of all through plaster casts and then at the life class. But it was in the profitable genre of topographical watercolour that he saw his future, and his first exhibit at the Academy was *The Archbishop's Palace, Lambeth* (Indianapolis Museum of Art), a watercolour shown there in 1790.

However, Turner's reputation was to be founded upon marine painting. His all-important first work in oil to

Fig. 8
Nicolas Poussin
The Deluge, 1660–4, Musée du Louvre, Paris

be shown there was *Fishermen at Sea*, exhibited in 1796 (fig. 1). Even at the time this was described as masterly, and over the following years his renown grew from a series of dramatic sea-pieces that he showed regularly. These included such *tours de force* as *Dutch Boats in a Gale: Fishermen endeavouring to put their Fish on Board*, exhibited in 1801, which became 'Picture of the Year' (fig. 7). Commissioned by the 3rd Duke of Bridgewater, as a pendant to his *Kaag close hauled in a Fresh Breeze* by Willem van de Velde the Younger (fig. 40 see p. 64), it simultaneously linked the young Turner with traditional marine painting and allowed him to challenge the dramatic and atmospheric effects of his Dutch predecessor. 'The Bridgewater Sea-piece' was followed in 1802 by 'The Egremont Sea-piece', painted for the 3rd Earl of Egremont, who was to become one of the artist's most important patrons (fig. 37 see p. 46).

In July 1802, like many other artists, Turner took advantage of the short-lived Peace of Amiens with Revolutionary France to make his first continental visit. Over about three months he toured the Swiss Alps and stayed in Paris. In the Louvre, opened as a public museum in 1793 and swollen with masterpieces requisitioned from all over Europe, he particularly admired Titian and Poussin, whose *Deluge* he described in his sketchbook with approving comments on its sublime colour (fig. 8).[12] It was to be the inspiration for his own painting of that title (fig. 5). Even more influential, however, was his turbulent Channel crossing, which led to the large seascape, *Calais Pier*, exhibited at the Royal Academy in 1803 (fig. 9). This thoroughly Romantic work depicts a packet boat crowded with passengers arriving in Calais amid stormy seas. It was based on his own experience, overlaid with reminiscences of similar works by Jacob van

Fig. 9
Calais Pier: An English Packet arriving, RA 1803
National Gallery, London

Ruisdael and his uncle Salomon. The danger of the situation and the fear of the travellers is communicated from the subject matter through Turner's very brushwork and embodied in his scintillating facture. In later years he claimed to have had himself bound like Ulysses to a mast during a storm so as to grasp the full force of the experience.[13]

This high drama also marked a milestone in the development of British marine painting. Whereas conventional depictions of the sea had favoured a pale, glossy, translucent swell whatever the weather, *Calais Pier* shows the sea as dark and heavy with streaks and outbursts of white foam dashed over its surface, all laid on thickly with a palette knife. Turner's contemporaries were amazed – but in the short term his bravura was to be his undoing. Reaction to *Calais Pier* heralds the beginning of critical disapproval of his works, which over the ensuing years were bitterly attacked by Sir George Beaumont and other arbiters of taste. The portraitist and prominent academician John Hoppner, who had accused Turner in 1798 of 'being a timid man, afraid to venture',[14] thought the treatment of *Calais Pier* and other works exhibited in 1803 'presumptive' and 'careless'.[15] It was generally condemned as unfinished and crude and Turner was particularly criticized for his use of pure white and reckless impasto. Sixty years later Walter Thornbury, an early biographer, probably revealed the artist's real motives for painting so boldly when he wrote: 'Anxious to avoid being too transparent and too slight in manner, his early oil-pictures were dark and heavy.'[16]

Alongside large, dramatic masterpieces Turner celebrated his love affair with the sea in other, quieter modes. They stemmed from his habit of touring the country in search of picturesque or antiquarian subject-

Fig. 10

Seascape with Squall coming up, c. 1803–4
Tokyo Fuji Art Museum, Japan

matter for his watercolours. He began with a tour to Bristol in 1791. In about 1793 he painted *A Storm off Dover* (cat. 10) and his sketchbooks used over the following years contain numerous marine subjects. In the early 1800s he executed a few small works in oils such as *Seascape with Squall coming up* (fig. 10). The Barber *Sun rising through Vapour* (cat. 7) falls between these and Turner's standard seascapes, whose size became established at three feet by four feet, the dimensions of *Fishermen at Sea* (see cats 2 & 9). Major works designed to steal attention at public exhibitions such as the National Gallery *Sun rising through Vapour* (cat. 6) were always much bigger.

Fig. 11
John Constable
His Majesty's Ship, Victory, Capt. H. Harvey, in the Memorable Battle of Trafalgar, between two French Ships of the Line, RA 1806, Victoria and Albert Museum, London

In 1805 the Battle of Trafalgar provided a focus for marine artists, and no event at sea so stimulated Turner's imagination. On 21 October that year the British fleet under Admiral Lord Nelson defeated the combined fleets of France and Spain off Cape Trafalgar near the Straits of Gibraltar, securing command of the seas for Britain during the rest of the Napoleonic Wars. At the moment of triumph Nelson himself was killed, making the event a cause of both national celebration and universal mourning. Nelson's body was brought home in his flagship, *Victory,* and lay in state at Greenwich, until his public funeral and burial in St Paul's Cathedral on 9 January 1806. Turner made a special journey to witness and record the *Victory's* arrival in the Medway in December 1805. She must have been an impressive sight in new black and yellow livery despite the battering she had taken. Ordered in 1758, the same year as Nelson's

birth, she had required 27 miles of rigging, four acres of sail and six years to build. By 1803 she was already famous as the veteran flagship of great triumphs at Ushant and Cape St Vincent and of distinguished admirals including Keppel, Howe and Hood. In April that year she was visited at Chatham dockyards by John Constable, who was greatly inspired by her beauty and drew her from three angles. These drawings were subsequently used in the preparation of his watercolour, *His Majesty's Ship, 'Victory', Capt. H. Harvey, in the Memorable Battle of Trafalgar, between two French Ships of the Line* (fig. 11). Turner also made a number of drawings, which formed the basis for several paintings including *The Battle of Trafalgar as seen from the Mizen Starboard Shrouds of the 'Victory'* (fig. 32 see p. 32) and *Portrait of the 'Victory' in Three Positions* (cat. 3). In the years following Trafalgar there was a massive demand for such patriotic works, much of it satisfied by the marine artists of the day such as Philippe Jacques de Loutherbourg (1740–1812) and Nicholas Pocock (cat. 19). In 1821 Turner himself returned to the subject with *The Battle of Trafalgar, 21 October 1805* painted for King George IV, in which he combined at close range several incidents that took place during the battle, creating an overwhelming image of death and destruction (fig. 12).

The years after 1805 were particularly significant for Turner's artistic development as he then kept a boat at Richmond; this may have been borrowed but was more probably built specially for him. In it he explored the Thames up and downriver, as well as its tributary, the Wey. Oil sketches on both panel and canvas were produced out of doors (cats 4 & 5), leading to a number of finished pictures, displayed at his new gallery in Harley Street over the next three years. These included both versions of *The Sun rising through Vapour* as well as *Sheerness as seen from the Nore* (cat. 8), praised together with another in 1808 by the critic of *The Examiner* as 'the finest sea-pieces ever painted by a British Artist.'[17] Approval was not universal, however, and Benjamin West told Farington in 1807 that he was disgusted with what he had seen at Turner's Gallery: 'views on the Thames, crude blotches, nothing could be more vicious.'[18] Later in life he acquired precise models of ships, with backgrounds and plaster waves that he himself seems to have painted with an appropriate bravura (fig. 13). It is not known for which pictures (if any) he used these, but ships were the height of military and commercial technology and important ones often had a uniquely recognisable shape and decoration, making accurate representations always essential.

The tradition of painting the sea and ships was very well established in Britain by the late eighteenth century.

Fig. 12
The Battle of Trafalgar, 21 October 1805, 1821
National Maritime Museum, Greenwich

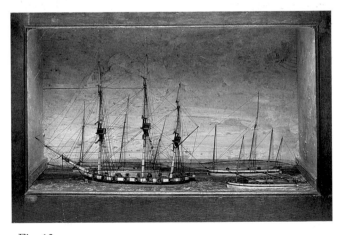

Fig. 13
Case containing Model Ships with Sea and Background painted by Turner, Tate Britain

Fig. 14
Simon de Vlieger
The Beach at Scheveningen, 1633
National Maritime Museum, Greenwich

Originating in the Netherlands it had grown rapidly in popularity there throughout the seventeenth century, reflecting the spread of the Dutch empire in the Indies and elsewhere. The first major practitioner was Hendrik Cornelisz Vroom (1566–1640), who used a palette of bright colours essentially derived from Pieter Bruegel the Elder. Jan Porcellis (before 1584–1632) was taught by Vroom but became a pioneer of the new Dutch realist school, moving towards a range of subdued tones more consonant with the variable skies of Holland. This idiom was further developed by his son Julius (cat. 16) and later Dutch painters, including Simon de Vlieger (fig. 14), Jan van Goyen and Ludolf Bakhuyzen. De Vlieger's monochrome style of the 1630s influenced Jan van de Cappelle and Willem van de Velde the Elder. It

was brought to England by Van de Velde, who came with his son in the winter of 1672–3. From then on Dutch pictures of the sea and shipping, especially those by Willem van de Velde the Younger greatly stimulated British interest in maritime art (cat. 17).

The paintings of the Van de Veldes were widely imitated, for example by Peter Monamy (1681–1749) who decorated supper-boxes in Vauxhall Gardens with marine pieces. Samuel Scott (1701/2–72) also owned pictures by Van de Velde and was heavily indebted to him in his early years before coming under the influence of Canaletto. Charles Brooking (cat. 18), Dominic Serres (1722–93) and Nicholas Pocock (cat. 19) all copied compositions by Van de Velde the Younger. His influence was further disseminated by the publication around 1725 of a series of sixteen mezzotints based on their sea-pieces. These were engraved by Elisha Kirkall and were particularly distinctive as they were printed in appropriate sea-green ink (cat. 22). Turner knew works by all these painters, or at least engravings after them and, as noted above, one of his first marine successes, 'The Bridgewater Sea-piece' (see cat. 26), was painted as pendant to a picture by Van de Velde the Younger (fig. 40). He had seen works by Van de Velde in the collection of John Hoppner and in later life recalled how influential a print after Van de Velde (cat. 22) had been for him with the words: 'This made me a painter'.[19]

From the end of the seventeenth century many ship portraits were produced reflecting increased British naval power, colonial expansion and prosperity dependent on maritime trade. A school of marine artists developed in the vicinity of the naval dockyards at Deptford. One of these, Dominic Serres, was well enough regarded to become a founder member of the

Fig. 15
George Morland
A Shipwreck, 1792
Private Scottish Collection

Fig. 16
Thomas Gainsborough
Coastal Scene with Sailors and Rowing Boats and Figures
on the Shore, c. 1780, Victoria and Albert Museum, London

Royal Academy in 1768 and went on to be appointed Marine Painter to King George III in 1780. Charles Brooking (c. 1723–59), born in Deptford, was perhaps the most talented of his generation and contributed considerably to the emergence of a national style in the 1750s (see cat. 18). However, to site Turner solely within the British marine school would be to give him far too restrictive a pedigree. This was recognised at the time by the critic of *The Morning Star*, who observed of 'The Mildmay Sea-piece': 'He seems to view Nature and her operations with a peculiar vision, and that singularity of perception so adroit that it enables him to give a transparency and modulation to the sea, more perfect than is usually seen on canvas:– he has a grace and boldness in the disposition of his tints and handling,

which sweetly deceives the sense; and we are inclined to approve him the more, as all our marine Painters have too servilely followed the steps of each other, and given us pictures more like Japanned tea-boards, with boats on a smooth and glossy surface, than adequate representation of that inconsistent, boisterous, and ever changing element.'[20]

Turner's artistic roots lay also with those landscapists who occasionally ventured into seascape, such as Thomas Gainsborough (1727–88), Julius Caesar Ibbetson (1759–1817) and George Morland (1763–1804). In the early 1780s Gainsborough painted about ten marines inspired by the East Anglian coastline. Morland diversified from village scenes in 1799 when he spent some time on the Isle of Wight in order to escape his

Fig. 17
School of C.-J. Vernet
Seaport at Sunrise, Dulwich Picture Gallery, London

creditors (fig. 15). For these artists, seascape provided a new context in which to situate picturesque staffage such as fishermen, wreckers or smugglers. Some of their works must have been known to Turner, although he was too young to remember Gainsborough's marines at the Academy in 1781 and 1783. However, one was owned by Earl Grosvenor, who made it available to students, and another, painted on glass, was sold by Margaret Gainsborough to Dr Thomas Monro, the young Turner's patron (fig. 16). Prints after drawings by Gainsborough, including several coast scenes, were published between 1802 and 1805.[21] Turner had copied some of Morland's genre scenes involving shipwrecks, fisherfolk, fish-markets and wreckers when he was only thirteen.[22]

More influential, however, were the storms and calms of Claude-Joseph Vernet (1714–89) and the sublime imaginings of de Loutherbourg. The leading painter of ports and the sea in France by the mid-1750s, Vernet

eventually became the most important maritime painter in Europe. From 1733 to 1752 he was in Italy, where he knew Richard Wilson (1713/14–82), an important formative influence on the young Turner. Vernet was hailed by Diderot, the influential French critic, as a greater artist than Claude and was commissioned by Louis XV to paint a series of the major commercial and military sea-ports of France. These may have been seen by Turner when he visited France in 1802; some at least must have been known to him through engravings. Numerous paintings by or attributed to Vernet were in British collections by 1800 (see cat. 23). *Seaport at Sunrise*, a work showing much in common with the Barber *Sun rising through Vapour*, was probably sold in London in 1795 (fig. 17).[23] About 1797 Turner made a memorandum of a Mediterranean coastal scene with a ship being careened, apparently based on a composition by Vernet.[24] Twelve years later he was compared with his predecessor by the critic 'Anthony Pasquin' (John Williams), who wrote of his *Spithead: Boat's Crew rescuing an Anchor*: 'The transparency of the water in this view is beautifully managed; and Turner may become, if he pleases, the first marine painter in the world; he has more correctness, though less brilliancy, than Vernet.'[25]

De Loutherbourg arrived from France in 1771 and exhibited at the Royal Academy from 1772 to 1812.[26] He established a reputation for storms, shipwrecks, *banditti* and set battles. Taken up by David Garrick at Drury Lane, he became a leading European stage designer. Many of his scenes used moonlight, with a rising or setting sun, or sunbeams piercing transparent trees (cf. cat. 1). Such imagery reappeared in his 'Eidophusikon', a miniature special effects show and popular attraction in Leicester Square from 1781 to 1793, visited by artists including Gainsborough, Reynolds and perhaps Turner himself. His *pièce de résistance* was that *locus classicus* of Burkean sublimity, the representation of a storm at sea – a direct ancestor of Turner's turbulent sea-pieces such as *The Shipwreck* (fig. 6). De Loutherbourg travelled extensively round the British Isles, producing seaside studies of fisherfolk and coastal views of Brighton, Ramsgate, Margate and other towns. In 1794 he went to Portsmouth with James Gillray to sketch men-of-war and he possessed a collection of ship models, similar to those owned by Turner, who was on familiar terms with him from about 1790 and must have known his *Deluge*, exhibited that year at Thomas Macklin's annual exhibition (fig. 18). Published as a print in 1797, it was prophetic of his own *Deluge* of 1805 (fig. 5).[27] In 1801 de Loutherbourg published *The Picturesque Scenery of Great Britain*, which featured a number of maritime views engraved in aquatint and anticipated by over a decade the appearance of W.B. Cooke's *Picturesque Views on the Southern Coast of England*, half of whose 80 prints were provided by Turner. In 1805 came de Loutherbourg's *The Romantic and Picturesque Scenery of England and Wales*, which included engravings and descriptions of a storm off Margate and the Needles, Isle of Wight.

Turner was even more deeply aware of a driving need to emulate the most august Old Masters. Few marine painters escaped his attention but his greatest interest lay in Claude Lorrain (1604/5–82) and Aelbert Cuyp (1620–91). Landscapes by Claude were among the most popular of those collected by Grand Tourists in the eighteenth century and there were plenty in British private collections – some of them accessible to visitors.

Fig. 18
Philippe Jacques de Loutherbourg
The Deluge, 1789–90, Victoria and Albert Museum, London

His popularity is confirmed by the increasingly high prices his works fetched in sales of the period, e.g. the two 'Altieri' Claudes.[28] They were seen at least twice by the young Turner when displayed by William Beckford in his London house in May 1799.[29] Between 1803 and 1805 Claude's *Seaport* and several other works formerly in European collections were acquired by John Julius Angerstein, at whose house in Pall Mall they were seen

Fig. 19
Claude Lorrain
Seaport, 1639, National
Gallery, London

by Turner (fig. 19). He would also have been aware of the recently published reproductions after Claude's drawings from the famous *Liber Veritatis*. The original was probably at Devonshire House in London for most of Turner's life, but in 1777 Richard Earlom brought it to the attention of a wider public with his two-volume collection of mezzotints. Turner's response was emphatic and unwavering. As early as 1799 he exhibited a watercolour of Caernarvon Castle, with a sunlight effect that can only have been derived from Claude.[30] In a lecture at the Royal Academy in 1811 he put something of his enthusiasm for Claude into words with a patriotic

fervour: 'Pure as Italian air, calm, beautiful and serene springs forward the works and with them the name of Claude Lorrain. The golden orient or the amber-coloured ether [...] replete with all the aerial qualities of distance, aerial lights, aerial colour [...] in no country as in England can the merits of Claude be so justly appreciated, for the choicest of his work are with us, and may they always remain with us in this country.'[31] His calm seascapes are unimaginable without the atmospheric sea-ports of Claude and it is no accident that in his second will of 1831 Turner specified that his first *Sun rising through Vapour* (cat. 6) should hang in the

Fig. 20
Claude Lorrain
Seaport with the Embarkation of the Queen of Sheba, 1648
National Gallery, London

Fig. 21
Aelbert Cuyp
The Maas at Dordrecht, BI 1815
Iveagh Bequest, Kenwood House, London

National Gallery between two paintings by Claude, including *Seaport with the Embarkation of the Queen of Sheba* (fig. 20).

Cuyp, sometimes known as the Dutch Claude, was being rediscovered by British collectors just when Turner was painting his early seascapes. Among his greatest pictures are the views of rivers and towns in the early morning or evening sun (fig. 21). Cuyp developed the new effect of *contre-jour* by moving his light source to a diagonal position at the back of his pictures and pioneered attempts to convey the full chromatic scale of sunrise and sunset. British painters, including Wilson and occasionally Gainsborough, strove to emulate his warm, golden tonality and the soft effects of his aerial perspective. In the last three decades of the eighteenth century a number of his works were engraved, and between 1800 and 1810 there

followed a period of speculation in his paintings, which appeared in a number of important sales.[32] Turner was thus made aware of such influential waterscapes as *River Scene with a View of Dordrecht*, which was bought by the Earl of Yarmouth in 1811 (fig. 22). His treatment of light and composition in both versions of *The Sun rising through Vapour* owes as much to Cuyp as to Claude. Ten years later he went on to pay his most explicit homage in *Dort or Dordrecht – The Dort Packet Boat from Rotterdam Becalmed*, exhibited at the Royal Academy in 1818 (fig. 24).[33]

It is also remarkable how the style of many of the figures in Turner's early seascapes was influenced by David Teniers the Younger (1610–90). Turner could have seen works by this Flemish painter in the collection of the Duke of Bridgewater, who had bought a number from the Orleans collection; and in the collection of Benjamin West, who owned four that would have been accessible to him. Teniers was also much engraved.[34] De Loutherbourg, who was responsive to Teniers, probably brought him to Turner's attention. In works such as the National Gallery *Sun rising through Vapour* (cat. 6) the influence of Teniers can be discerned in the individuality of the figures and their raw, caricature-like appearance (fig. 23). Turner devoted a whole section to Teniers in his influential 'Backgrounds' lecture delivered at the Royal Academy in 1811, giving him more attention than Cuyp, for example. The young David Wilkie, freshly arrived from Scotland, caused a sensation at the Royal Academy exhibition of 1806 with his Teniers-like *Village Politicians,* but Turner's interest seems to pre-date Wilkie's on the evidence of his early seascapes.[35]

By 1805 Turner was recognised as the leading young painter of the day, with a reputation based largely on seascapes, confirmed by the exhibition of *The Shipwreck* and its publication in mezzotint (cat. 24). It was at this point that he devised the *Liber Studiorum* (Book of Studies) as a way of establishing an academic, theoretical framework for the various kinds of landscape. Conceived as a visual treatise on landscape art, it is the central document of Turner, the theorist of painting, who intended it to demonstrate the range and improve the status of landscape from its traditionally low ranking in the hierarchy of genres. In title and medium it imitated Earlom's famous mezzotints after Claude's *Liber Veritatis*. Rather than produce a mere record, however,

Fig. 22
Aelbert Cuyp
River Scene with a View of Dordrecht, c. 1647
Wallace Collection, London

Turner aimed to illustrate the various 'branches' into which landscape painting could be divided, together with the range of his own abilities. He identified six landscape categories: Historical, Mountainous, Pastoral, Marine, Architectural and 'E.P.', generally believed to signify Epic Pastoral. The *Liber Studiorum* was published in fourteen parts plus frontispiece from 11 June 1807 to 1 January 1819 (see cats 12 and 14). With one exception, each part comprised five prints, making 71 in all, of which nine were designated as marine.[36]

With the completion of *The Wreck of a Transport Ship* in 1811 or 1812, Turner's first series of seascapes reached a fitting climax (fig. 2). Sometimes known as *The Wreck of the Minotaur on Haak Sands*, referring to a disaster that took place in December 1810, this 'most splendid sea piece that has ever been painted'[37] develops even further the horrifying tumult of *The Shipwreck*, painted about five years earlier (cf. cat. 24). Turner did not cease to paint seascapes altogether after this but exhibited major examples far less regularly over the ensuing years.

Fig. 24

Dort or Dordrecht: The Dort Packet-Boat from Rotterdam Becalmed, RA 1818, Yale Center for British Art, Paul Mellon Collection, New Haven

The greatest exception occurred in 1818, when he showed the dazzling *Dort or Dordrecht* at the Royal Academy (fig. 24). This was inspired by his first visit to Holland and Germany in 1817 and owed an obvious debt to Cuyp in style and subject-matter. It may be seen as a watershed in Turner's art – simultaneously the culmination of his study of the Old Masters and the prelude to the subtlety and brilliance of the highly personalised visions of his later career. Appropriately, it was bought by his friend and patron, Walter Fawkes who also owned the Barber *Sun rising through Vapour*.

Not until his 1827 visit to John Nash at East Cowes on the Isle of Wight was Turner's engagement with the sea revived. Stemming from this came another great series of

Fig. 23

David Teniers the Younger

Kermesse, 1648, Staatliche Kunsthalle, Karlsruhe

Fig. 25
The Wreck Buoy, c. 1807/reworked 1849,
Walker Art Gallery, Liverpool

Fig. 26
Richard Parkes Bonington
A Sea Piece, 1824–5, Wallace Collection, London

sea-pieces over the 1830s and 1840s, which developed the range of his earlier marines to include Venetian scenes, and slaving and whaling subjects. These were couched in a vaporous style that often baffled patrons and alienated critics. Something of the contrast with his early sea-pieces is embodied in a single painting, *The Wreck Buoy* (fig. 25). Painted about 1807, the picture was sent for exhibition in London in 1849. On seeing it again Turner was dissatisfied and insisted on reworking it, heightening the colour of the sails of two of the boats, highlighting some of the wave crests and inserting a scintillating rainbow. His blushing red additions to the sky, made over a period of six days to the alarm of its owner, H.A.J. Munro of Novar, embody his later manner and mix uncomfortably with the darker, heavier layers of original paint. *The Wreck Buoy* may be taken to enshrine Turner's lifelong interest in the sea and the depiction of the sublime, overlaid with musings on the frustration of all human hopes and the inevitability of death.

While Turner's late seascapes had no imitators, his early works were highly influential and inspired Augustus Wall Callcott (cat. 20), Clarkson Stanfield (1793–1867), John Crome (1768–1821), William Daniell (1769–1837)[38] and Richard Parkes Bonington (1802–28). Born in England, Bonington moved to France in 1817, where he received his early teaching in Calais from Louis Francia (cat. 21). In the mid-1790s Francia had worked as a copyist of John Robert Cozens and other watercolourists with Turner and Girtin at Dr Monro's evening 'Academy' in the Adelphi. He thus forms a direct link in the succession of watercolour seascapes stretching from Cozens to Bonington. Before 1825 Bonington himself could have known Turner's works only through engravings. When he first came to London that year he spoke much of the wonder of Turner, and his *Sea Piece* (fig. 26) certainly recalls Turner's *Portrait of the 'Victory' in Three Positions* (cat. 3) as well as works by Francia.[39] His first oils included the ambitious *Fishmarket near Boulogne* (fig. 27), one of five works that secured a gold medal for him at the Paris Salon of 1824. After his premature death in 1828 many of Bonington's works were sold in London and Turner seems to have known these and taken ideas in turn from the younger artist. Bonington was not alone in making a very important contribution to the development of marine art in watercolour, for in the first three decades of the nineteenth century several major watercolourists periodically turned to seascape, of whom the best were John Sell Cotman (1782–1842) and the Birmingham artist, David Cox (1783–1859). Between 1805 and 1810, and again around 1820, Cotman produced some of his

Fig. 28
John Sell Cotman
The Dismasted Brig, c. 1823, British Museum, London

Fig. 27
Richard Parkes Bonington
Fishmarket near Boulogne, Salon 1824, Yale Center for British Art, Paul Mellon Collection, New Haven

Fig. 29
David Cox
The Entrance to Calais Harbour, c. 1829 (?), Yale Center for British Art, Paul Mellon Collection, New Haven

Fig. 30
Samuel Owen
Rough Weather: Dutch and English Fishing Boats, 1805
Birmingham Museums and Art Gallery

greatest works, including *The Dismasted Brig* (fig. 28), while a few years later Cox executed his serene *Entrance to Calais Harbour* (fig. 29). Other major seascapists included Samuel Prout (1783–1852), Samuel Owen (fig. 30) and Anthony Vandyke Copley Fielding (1787–1855).

Turner's preoccupation with light as the one visual reality is the constant that links his early and late styles and pervades his seascapes of all kinds. Throughout his career he produced hundreds of works devoted to the study of sunlight and the effects of light on sea and land, uniquely recording dramatic or beautiful sunrises and sunsets in a variety of media. Both versions of *The Sun rising through Vapour* evoke atmospheric qualities of ineffable serenity that look forward to his aerial visions of the 1830s and 1840s. From the beginning Turner was fascinated by the immaterial vehicles of palpable reality, such as steam, smoke and mist. As he stood at the threshold of a successful career as painter, printmaker, academician and professor, these two tranquil marines act as a metaphor for the ambitious young artist's own sun-like ascent from the mists of obscurity towards a golden future of security and prosperity.

1 John Burnet, *Turner and his Works*, London, 1852, p. 75.
2 E.T. Cook and A. Wedderburn, eds, *The Works of John Ruskin*, XXXV, London, 1908, p. 571 (*Dilecta*, chapter 1).
3 According to the engraver Edward Bell's manuscript notes cited by W. Thornbury, *The Life of J.M.W. Turner, R.A.*, London, 1862, I, p. 75.
4 L. Herrmann, 'Turner and the Sea', *Turner Studies*, I, no. 1, [1981], p. 4.
5 Introduction to Sir J.G. Dalyell's *Shipwrecks and Disasters at Sea*, Edinburgh, 1812, quoted by T.S.R. Boase, *Journal of the Warburg and Courtauld Institutes*, XXII, 1959, p. 332.
6 Samuel Arnold's opera, *The Shipwreck*, was performed at Drury Lane in 1796.
7 Its full title was *Naufragia: or Historical memoirs of Shipwrecks and of the Providential Deliverance of Vessels*. A more extensive publication was Archibald Duncan's six-volume *The Mariner's Chronicle, Being a Collection of Fires, Famines and Other Calamities Incident to a Life of Maritime Enterprise*, London, 1810.
8 Cantos 1 and 2 of Byron's *Don Juan* containing graphic accounts of Juan's ordeal at sea and the wreck of *Trinidada* were published in 1819.
9 *The Times*, 3 May 1797.
10 Thornbury, *op. cit.*, 1862, I, p. 302.
11 Thornbury, *op. cit.*, 1862, I, p. 25.
12 'Studies in the Louvre' Sketchbook, 1802, TB LXXII, pp. 41a and 42.
13 On the genesis of this myth see also G. Levitine, 'Vernet tied to a mast in a storm: The evolution of an episode of art historical Romantic folklore', *Art Bulletin*, XLIX, 1967, pp. 93–100.
14 K. Garlick and A. Macintyre eds, *The Diary of Joseph Farington*, III, New Haven and London, 1979, p. 963 (5 January 1798).
15 K. Garlick and A. Macintyre eds, *The Diary of Joseph Farington*, VI, New Haven and London, 1979, p. 2021 (30 April 1807).
16 Thornbury, *op. cit.*, 1862, I, p. 258.
17 *The Examiner*, 8 May 1808.
18 K. Cave, ed., *The Diary of Joseph Farington*, VIII, New Haven and London, 1982, p. 3038 (5 May 1807).
19 Thornbury, *op. cit.*, revised edition 1877, p. 8. For the influence of Hoppner's *A Gale of Wind* and other seascapes on the young Turner see J.H. Wilson, 'The Landscape paintings of John Hoppner', *Turner Studies* VII, no. 1, 1987, pp. 15–25.
20 Cited in M. Butlin and E. Joll, *The Paintings of J.M.W. Turner*, New Haven and London, 1984, I, p. 3.
21 W.F. Wells and J. Laporte, *A Collection of Prints illustrative of English Scenery from the Drawings and Sketches of Gainsborough*, London, 1802–5.
22 Thornbury, *op. cit.*, 1862, I, p. 258.
23 With the Noel Desenfans Collection. In 1811 it was bequeathed to Dulwich College by Sir Francis Bourgeois, a fellow Royal Academician of Turner's.
24 'Wilson' Sketchbook, TB XXXVII, p. 105.
25 *Morning Herald*, 4 May 1809.
26 See further R. Joppien, ed., *Philippe Jacques de Loutherbourg, RA, 1740–1812*, exh. cat., London, Kenwood House, 1973.
27 Deluges and shipwrecks went on to become favourite subjects for Romantic artists, including John Martin and Francis Danby.
28 *Landscape with the Father of Psyche sacrificing at the Milesian Temple of Apollo* (1663) and *Landscape with the Landing of Aeneas in Latium* (1675; both The National Trust, Fairhaven Collection, Anglesey Abbey).
29 K. Garlick and A. Macintyre, eds, *The Diary of Joseph Farington*, IV, New Haven and London, 1979, pp. 1219–20 (8 and 9 May 1799).
30 M. Kitson, 'Turner and Claude', *Turner Studies*, II, no. 2, 1983, p. 5.
31 J. Ziff, '"Backgrounds, Introduction of Architecture and Landscape": A Lecture by J.M.W. Turner', *Journal of the Warburg and Courtauld Institutes*, XXVI, 1963, pp. 144–5.
32 S. Reiss, *Aelbert Cuyp*, Boston, 1975, p. 11.
33 Cuyp's *The Maas at Dordrecht, Evening* (fig. 21) was in an exhibition at the British Institution in 1815. It influenced not only Turner but also Callcott, who exhibited a highly Cuypian *Entrance to the Pool of London* (Bowood House) at the Royal Academy in 1816.
34 James Holworthy, a close friend and confidant whom Turner knew by 1805, owned over 200 prints after Teniers the Younger.
35 E. Shanes, *Turner's Human Landscape*, London, 1990, pp. 314–323. The author is grateful to Eric Shanes for drawing this important reference to his attention.
36 The last marine subject was *Entrance of Calais Harbour*, published on 1 January 1816.
37 W.L. Wyllie, *The Life of J.M.W. Turner*, London, 1905, p. 48. Wyllie was a leading marine painter of his day as well as a penetrating critic of Turner's seascapes.
38 K. Cave, ed., *The Diary of Joseph Farington*, VIII, 1982, p. 2808 (11 July 1806).
39 J. Ingamells, *The Wallace Collection, Catalogue of Pictures I, British, German, Italian, Spanish*, London, 1985, p. 22.

CATALOGUE

TURNER: PAINTINGS

1. *Moonlight, a Study at Millbank*, RA 1797
Oil on mahogany, 31.5 x 40.5 cm
Tate Britain

Exhibited when Turner was only 22 years old, the view is of the Thames, looking east. The work is in the tradition of moonlight scenes by C.J. Vernet, Joseph Wright of Derby (fig. 31) and P. J. de Loutherbourg, especially in its contrast of the cold, natural light of the moon with the ruddier glow of lantern light. Turner may also have had in mind the nocturnes of the Dutch landscapist Aert van der Neer (1603/4–77), who specialized in moonlight effects. Such subjects were a fashionable pictorial convention by the late eighteenth century and popular with British collectors. In 1796–7 Turner also executed a small group of nocturnal scenes in watercolour, including *Moonlight over the Sea at Brighton* (Tate Britain), where the moon is partly obscured by clouds, allowing light to be reflected from the waves in the foreground.

The present study, which may have been a technical exercise connected with *Fishermen at Sea* (fig. 1 see p. 7), nevertheless gives an impression of uncontrived naturalism. Its richness partly depends upon the smooth quality achieved through painting on panel, the support generally favoured by Van der Neer. There are compositional similarities to the Barber *Sun rising through Vapour,* notably the position of the celestial disc and of the ship silhouetted at the right, while the pervasive nocturnal calm also anticipates the dawn mood of the later picture. A little to the west of the spot from where the view was taken stood the house at 6, Davis Place in which Turner was to die over 50 years later.

Fig. 31
Joseph Wright of Derby,
Moonlight, Coast of Tuscany, RA 1790,
Tate Britain

2. Fishermen upon a Lee Shore in Squally Weather
RA 1802
Oil on canvas, 91.5 x 122 cm
Southampton City Art Gallery

This dramatic sea-piece was exhibited at the Royal Academy in the year Turner was elected an academician. Its wildness and darkness, consonant with contemporary enthusiasm for the aesthetic of the sublime, concentrate and refine the power of larger seascapes such as *Calais Pier* (fig. 9). A keen fisherman himself, Turner was fascinated by the heroism of those who went to sea, a group of whom he shows here struggling with their vessel in the teeth of heavy winds. Boats in the background and the distant jetty provide a pictorial framework for the scene but the prominent anchor – a traditional symbol of hope – serves to emphasize the fastness of the shore in contrast to the agitated waves. In front of it a mooring ring and scattered mussels indicate

Turner's feel for the minutiae of the sea, while the gliding gulls trace almost tangible air-currents.

The mood of this work is in marked contrast to the golden serenity of both versions of *The Sun rising through Vapour*. It emulates the theatrical set-pieces of de Loutherbourg while the parting of the clouds to admit light recalls the manner of Wright of Derby (fig. 31). Turner thus transcends mere description of the fishermen's precarious existence by locating it within a larger scheme of nature, brilliantly capturing a mood of danger and mutability.

Nevertheless, the picture was not well received. The critic of the *True Briton* considered the waves 'chalky', the water lacking in transparency and the whole work 'much too indeterminate and wild.' The critic of the *Star* echoed these sentiments, calling it an 'admirable sketch' but doubting whether 'Mr Turner could himself make it a good finished picture.' Bought by the banker Samuel Dobree it passed in 1853 into the possession of Charles Birch, an active Birmingham collector.

3. *Portrait of the 'Victory' in Three Positions, passing the Needles, Isle of Wight*, c. 1806
Signed and dated (?) lower right: *JMW Turner RA 1806 (?)*
Oil on canvas, 67 x 100.3 cm
Yale Center for British Art, New Haven,
Paul Mellon Collection

The Battle of Trafalgar took place on 21 October 1805, confirming Britain's maritime supremacy and removing the threat of Napoleonic invasion. Nelson himself, however, was killed by a French marksman. His flagship

Fig. 32
The Battle of Trafalgar as seen from the Mizen Starboard Shrouds of the 'Victory', 1805, Tate Britain

returned with his body, anchoring off Sheerness in Kent on 22 December. Turner made a special journey to witness her arrival and drew her as she entered the Medway. Afterwards he executed a large number of detailed studies on board the ship.[1] This painting, however, is located off the Isle of Wight, with the Needles in the distance and the south-west coast running towards Freshwater Bay. If this orientation is taken literally, the ship is heading towards the south-west tip of the island rather than through the Solent. It has been questioned whether the ship really is the *Victory*. She is not easily identifiable as such (see cat. 19) and it is unlikely that, with Nelson's body on board, her ensign would be flying at full-mast.

By choosing the Isle of Wight as backdrop Turner may have been reacting to recent criticisms of its landscape by the Reverend William Gilpin. His 1798 guidebook of the island condemned the cliffs for being

'one long monotonous white.'[2] Yet it is their very whiteness that Turner, who visited in 1795, has chosen to emphasize. The cliffs were noticed more favourably by de Loutherbourg in his *Romantic and Picturesque Scenery of England and Wales*, published in 1805.

Portrait of the 'Victory' may have been shown in Turner's Gallery, Harley Street, together with the much larger *Battle of Trafalgar as seen from the Mizen Starboard Shrouds of the 'Victory'*, which Joseph Farington saw there in an unfinished state on 3 June 1806 (fig. 32). It was bought directly from Turner by Walter Fawkes, who later owned the Barber *Sun rising through Vapour* (see cat. 15). It is likely that the two works, of almost equal dimensions, were intended as pendants from an early stage.

[1] 'Nelson' Sketchbook, Tate Britain, TB LXXXIX. At Tate Britain there are also two larger studies taken of the deck, TB CXX-c and Vaughan Bequest CXXI-S, in addition to a monochrome watercolour study of *The 'Victory' coming up the Channel with the Body of Nelson* (TB CXVIII c).

[2] *Observations on the Western Parts of England, to which are added a few remarks on the Picturesque Beauties of the Isle of Wight*, London, 1798, p. 336.

4. *Coast Scene with Fishermen and Boats*, c. 1806–7
Oil on canvas, 85.7 x 116.2 cm
Tate Britain

This highly evocative oil sketch is probably set in the Thames estuary, an area of great significance to Turner in the years leading up to *The Sun rising through Vapour*. Fishermen are shown digging for bait in the centre foreground. To the right two boats are drawn up on the sand for repair with three attendant figures dimly discernible. In the lower right corner the outline of a large anchor is suggested. At left beyond the breaker are two half-rigged fishing boats and

further out to the right a man-of-war, possibly the guardship at Sheerness.

The work is one of a series of Thames sketches on canvas datable to these years, of which some at least are likely to have been begun out of doors. They are all painted over a dry chalky ground and roughly similar in size (see cat. 5). Here Turner's free technique and subtle colour harmonies anticipate the watercolours executed after his return from Italy in 1819, while his bold treatment of the sky looks forward to the liberation of light and colour in his later seascapes. Turner returned to the Thames estuary on numerous occasions, setting perhaps his most famous work, *The Fighting Téméraire* (National Gallery, London) there in 1838.

5. *Margate, setting Sun*, c. 1806–7
Oil on canvas, 85.7 x 116.2 cm
Tate Britain

Turner's links with Margate commenced in his early years. At the age of eleven he was sent there to stay with relatives of his mother and seems to have attended Mr Coleman's school. In 1796 he recorded the bay and harbour ('Wilson' Sketchbook, Tate Britain); later sketchbooks confirm his presence in 1806–8 and 1815–16. Although this painting is unfinished and not easy to interpret, it seems to show the stone pier on the left with the harbour to its right and Hooper's Mill on Fort Hill behind.[1] Turner had already shown a small view of *Old Margate Pier* at the Royal Academy of 1804 and he went on to exhibit a view of the town from further out to sea in 1808 (fig. 33). This inward-looking view underlines his fascination with the details of seaports and their commerce, anticipating, for example, *St Mawes in the Pilchard Season* (1812, Tate Britain). In 1824 a view of the

Fig. 33
Margate, 1808, Tate/Egremont Collection, Petworth House, Sussex

town was engraved for inclusion in W. B. Cooke's *Picturesque Views of the Southern Coast of England*.

During the opening decades of the nineteenth century, Margate rapidly developed from a quiet fishing town to a popular resort for visitors from London. These included artists such as Joshua Cristall and Philippe Jacques de Loutherbourg, who described it as 'a bathing place with which none can vie [having] … all the usual accompaniments … bathing machines, circulating libraries, toy-shops, card and assembly rooms, and a theatre.'[2] From about 1833 Turner spent much time in Margate at the house of Sophia Booth until she moved to Chelsea in 1846 to keep house for him there.

[1] E. Shanes, *Turner's Rivers, Harbours and Coasts*, London, 1981, pp. 26–7.
[2] De Loutherbourg, *The Romantic and Picturesque Scenery of England and Wales*, 1805 [unpaginated].

6. *The Sun rising through Vapour: Fishermen cleaning and selling Fish*, RA 1807
Oil on canvas, 134 x 179.5 cm
National Gallery, London

Fisherfolk unload their catch and prepare it for sale before a hazy sunrise over a calm, reflective sea, presenting a poignant contrast between peaceful nature and noisy humanity. The subject is compositionally related to the later work at the Barber Institute (cat. 7) but the jetty is more prominent and set on the right, while the warships are smaller and further out to sea. Their ghostly appearance results from the sea-mist or 'vapour' which has not yet been dispelled by the heat of the sun. Although it may be earlier in the day, the scene is busier than the Barber Institute's picture. Great attention has been paid to the objects on the seashore – fish, crabs, mussels and starfish – as well as the large anchor in the left foreground. All this is possibly the natural effect of the picture's large dimensions (increased during its execution) but there may also be symbolic reasons.

The boats and figures reflect the work of Dutch painters such as Cuyp, Bakhuyzen and Van de Cappelle, a debt confirmed by the picture's title, *Dutch Boats*, when hung in Turner's Gallery in 1810. However, Turner goes further than any of his predecessors by directly confronting the sun as the source of light. It is also remarkable how close some of his figures are to the style of David Teniers the Younger. The man wearing a flat, Flemish cap, with a hunched back seen from behind, is close to one in Teniers's *Kermesse* sold in London in 1794 (fig. 23).[1] These traits may in part be explained by Turner's desire to suggest a Dutch setting. Writing in 1819 William Carey stated: 'This scene is supposed to represent a harbour on the coast of Holland; although the artist has not confined himself to the particulars of a local view. The shipping are [sic] built like those which were used by the Dutch towards the close of the sixteenth century'.[2] Certainly the theme is that of a 'calm' rather than a view, a subject pioneered by Dutch marine painters (see cat. 17). Even so, the picture as a whole is also the most salient example of Turner's acute understanding of atmosphere, stemming from his *plein-air* studies made along the Thames over the preceding two years.

Following its appearance at the Royal Academy, the painting was shown at the British Institution in 1809 and in Turner's own gallery the following year. Five drawings in the Turner Bequest can be related to its composition. It was purchased by Sir John Leicester in 1818 for 350 guineas but bought back by Turner upon Leicester's death in 1827 (for a larger sum). Four years later, the artist specified in his second will that it should hang in the new National Gallery together with his *Dido building Carthage* between two paintings by Claude, who specialized in depictions of sea-ports at sunrise or sunset. These were *'The Mill'* and *Seaport with the Embarkation of the Queen of Sheba* (fig. 20).

[1] E. Shanes, *Turner's Human Landscape*, London, 1990, p. 314.
[2] W. Carey, *A Descriptive Catalogue of Paintings by British Artists in the Possession of Sir John Fleming Leicester, Bart*, 1819, p. 22.

7. *The Sun rising through Vapour*, c. 1809
Signed lower right: *J.M.W. Turner, R.A.*
Oil on canvas, 69 x 102 cm
The Barber Institute of Fine Arts,
The University of Birmingham

In the right foreground, a group of men and women display freshly caught fish for sale on the sand, while more are being unloaded from a boat moored behind them. At the far left of the group stands a fisherman in a red cap, holding an anchor and looking out to sea. Opposite, a group of sailing and rowing boats are drawn up next to the jetty onto which baskets of fish are being unloaded. Above the horizon the golden disc of the newly risen sun suffuses the sky with a shimmering hue and disperses banks of clouds. Before it is moored a

Fig. 34
Hastings, Fishermen on the Sands, TG 1810,
The Nelson-Atkins Museum of Art,
Kansas City, Missouri

first-rate man-of-war of 100 guns being approached by a small boat and partly obscuring a dismasted hulk and another sailing boat. A gun is fired from the hulk, perhaps to mark the hour or in salute to approaching dignitaries. Another sailing boat and a rigged frigate are visible further out to sea on the right. These warships act as potent symbols of Britain's sea-power, protecting the quiet but essential industry which underpins her commercial strength and self-sufficiency at a time of war and blockade by France.

Despite prototypes in the work of Claude, Cuyp, Wilson and Vernet, the subject of sunlight viewed through mist was deemed unworthy of landscape by most of Turner's contemporaries because the thick atmosphere which resulted prevented much introduction of colour. Here, however, the transfiguring power of light, together with its infinite subtlety is brilliantly observed in the miraculous hues of water and sky that greet the awakening dawn.

The sole pencil study is contained in the 'Spithead' Sketchbook (cat. 13). In the past the Barber picture has often been eclipsed by the larger and earlier painting of the same title in the National Gallery (cat. 6), yet it is not a replica as the two compositions differ significantly. Both works, however, combine the Anglo-Dutch tradition of 'shipping becalmed' or 'warships anchored off' with that of the beach fish-market, popularized by Simon de Vlieger and others (fig. 14, p. 18). Some of the

figures themselves have a noticeably Dutch appearance but there is no reason to suppose that the present setting is intended as anywhere but England, given the ensign displayed on the principal man-of-war. The picture was possibly one of two works entitled *Fishing-boats in a Calm* exhibited in Turner's Gallery in 1809.[1] Its title has varied over the years; in 1823, for example, it was recorded as *Calm: Three Deckers at Anchor*.[2] It was purchased from Turner by his friend and patron, Walter Fawkes of Farnley Hall, near Leeds, Yorkshire, and appears in Turner's watercolour of the drawing room at Farnley (cat. 15). Another similar picture, *Hastings, Fishermen on the Sands*, was exhibited in Turner's Gallery in 1810 (fig. 34).

[1] A.J. Finberg, *The Life of J.M.W. Turner, R.A.*, 1939, p. 41, nos.141, 144.
[2] In the exhibition of the Northern Society, Leeds, held that year. (D. Hill, *Turner's Birds*, Oxford, 1988, pp. 10, 11 and 26 [note 5].)

8. *Sheerness as seen from the Nore,* TG 1808
Oil on canvas, 104.5 x 149.5 cm
Richard Green, London

This magnificent work is set off Sheerness, one of four major naval dockyards on the Thames, along with Deptford, Woolwich and Chatham. Standing at the confluence with the Medway, the great naval and merchant shipping anchorage is seen here from the south east, at a point where the estuary is six miles wide. Turner records the scene from a boat riding low in the choppy water – probably similar to the ones depicted in the right foreground. In the distance at far left the sun has just begun to rise, scattering highlights across the sea and infusing the sky with a roseate glow. Before it the ghostly form of the Admiralty guardship is thrown into silhouette, while opposite the distant buildings and fortifications of Sheerness are spotlit. Nearer the viewer, fishermen, with a number of fish already to their credit, narrowly avoid collision with a sailing boat, thus adding to the natural drama of the seascape, whose heaving peaks and troughs are reflected in the volatile eddies of the threatening sky.

Given the wartime context and the Mutiny on the Nore of 1797, which had threatened the Royal Navy's command of the seas and jeopardized the commerce of London, the protective relationship between the guardship and the fishing-boats is a poignant one, recalling the similar relationship in the Barber *Sun rising through Vapour*. The picture was bought by Samuel Dobree (1759–1827), a city merchant and banker and a keen supporter of contemporary artists, who also owned three other seascapes by Turner, including *Fishermen upon a Lee Shore in Squally Weather* (cat. 2). A number of other views of Sheerness were exhibited by Turner in his gallery in 1808 and 1809 (see p. 69).

9. *Fishing upon the Blythe Sand, Tide setting in*, TG 1809
Oil on canvas, 88.9 x 119.4 cm
Tate Britain

Blythe Sands are in the Thames estuary off Sheerness, facing Canvey Island. An enthusiastic fisherman himself, Turner often depicted the activities of commercial fisherfolk, but here he concentrates on the daily routines of fishing and the mood of sea and sky to the virtual exclusion of other details. He pays particular attention to the ship in the centre, which he studied in a watercolour. By focusing on its dark sails he silhouettes it against the lighter ships behind, which indicate the brisk activity on this stretch of water. They bask in sunlight emanating from above the heavily scumbled, embanked clouds. The expanse of sky suggests a freshening wind accompanying the returning tide, a restlessness and change that are emphasized by the gulls gliding over the sand in the foreground.

The painting was shown in Turner's Harley Street Gallery. Each year potential patrons were invited to the opening of his exhibition there but despite a strong desire to sell his works Turner apparently turned down an offer from his critic, Sir George Beaumont, a leading figure of the art establishment. The following year the painting seems to have been offered to Sir John Fleming Leicester, a pioneer in the collecting of contemporary British art. Despite further exhibitions at Turner's Gallery in 1810 and in 1815 at the Royal Academy and in Plymouth, it remained unsold and Turner became particularly fond of it. However, in later years, when he became neglectful of his pictures, it 'served as the blind to a window that was the entrée of the painter's favourite cat.'[1]

[1] Thornbury, *op. cit.*, 1862, II, p. 173.

10. *A Storm off Dover*, 1793 (?)
Pencil and watercolour, 25.6 x 36.1 cm
Tate Britain

This is one of Turner's earliest depictions of the sea, probably dating from 1793, when he made a tour of Kent, including Dover. It shows how much he had already assimilated the technique of the topographical watercolour, but is overlaid with the effects of his own observations on the spot. Nevertheless, it seems to have been executed at the Monro Academy and may have been intended by Turner for other members to copy.[1]

While natural to such a subject, the greens, blues and greys of this work are firmly based in the palette of the preceding generation of watercolourists, such as Thomas Malton and Edward Dayes, while the pencil work owes much to Girtin. The sublime effect of the storm, however, is far more dramatic than anything that had gone before, its turbulence not only stirring up sea and sky but dislocating the scale and orientation of the town itself. Buildings and sea-front appear overwhelmed by the sheer volume of sea and the exaggerated size of the ship and flag. Even the white cliffs, a poignant symbol of the British nation, are obscured – perhaps an allusion to the increasing threat to Britain from Revolutionary France. The vulnerability of the town is further emphasized by its depiction in outline only.

By the early 1790s Turner was well aware of the shipwreck scenes of Vernet and de Loutherbourg (fig. 35); storms and disasters were to become staple subject-matter throughout his entire career (see cat. 24). This watercolour, however, is a very early example of such natural violence and man's powerlessness in the face of it.

Fig. 35
Philippe Jacques de Loutherbourg
Shipwreck, RA 1793, Southampton City Art Gallery

[1] A. Wilton, 'The "Monro School" Question: Some Answers', *Turner Studies*, IV, no. 2, 1984, p. 22.

11. *Fishermen lowering Sail; A Three-Master under Sail
in the Distance*, c. 1796
Gouache, pencil and watercolour, 27.2 x 35.6 cm
Tate Britain

This is another early example of Turner's interest in the daily life of fishermen. Seen from the shore three men are attempting to control their boat by lowering its sails as the wind picks up at the approach of a storm. The telling juxtaposition of small fishing vessels with ocean-going ships out to sea was a compositional device used in *The Sun rising through Vapour* and other later works. Here the three-master, dimly shadowed by a cutter behind, is seen as a ghostly presence smoothly progressing in the distance while the boat in the foreground struggles in rough water just beyond the beach (cf. cat. 2).

The subdued colours of the elements are brightened by the yellows and blues of the sailors' clothes and the light red lining of the boat. In tonal range, though less so in technique, the work has much in common with a group of coastal views executed by Thomas Girtin from about 1798. These were presumably based on drawings made during tours of the south west counties in 1797 (fig. 36) and the north east of England in 1800 and 1801. The present watercolour is probably connected with Turner's visit to Southampton and the Isle of Wight, which took place in August or September 1795. In his early sketchbooks, especially the 'Wilson' Sketchbook of 1797, there are a number of spirited sea studies that anticipate Turner's early marines in oil. He was one of the first artists to realize how admirably suited marine subjects were to the innate qualities of transparency, immediacy and atmospheric effect in watercolour (cf. cat. 21).

Fig. 36
Thomas Girtin
Lyme Regis, Dorset, c. 1797
Yale Center for British Art,
Paul Mellon Collection,
New Haven

44

Fig. 37
Ships bearing up for Anchorage
('The Egremont Sea-piece),
RA 1802, Petworth House, Sussex

12. *Ships in a Breeze*, c. 1806–7
Pen and ink, pencil and watercolour, 18 x 25.9 cm
Tate Britain

In 1802 Turner exhibited *Ships bearing up for Anchorage* at the Royal Academy (fig. 37). This was the earliest and probably the first painting of his to be bought by George Wyndham, 3rd Earl of Egremont, who was to become a major patron. The work became known as 'The Egremont Sea-piece'. Together with the slightly earlier 'Bridgewater Sea-piece' (see cat. 26) it marks the point at which Turner began to apply Poussinesque principles of structure and composition to his marines. He had first experimented with these concepts in a historical landscape, *The Fifth Plague of Egypt* (RA 1800, Indianapolis Museum of Art). In doing so, he transformed the Dutch seascape tradition that had remained static in England since the arrival of the Van de Veldes in the late seventeenth century.

Four years later Turner executed this highly atmospheric watercolour in preparation for publication in the *Liber Studiorum*. He condensed and simplified the composition of his painting, clarifying the design by emphasis on the selective fall of light, notably that articulating the group of overlapping ships in the centre. As in the Barber *Sun rising through Vapour*, the jetty to the left serves as a *repoussoir* to establish scale and distance in the work. The watercolour was etched by Turner himself and engraved in mezzotint by Charles Turner. The print then appeared in Part II of the *Liber Studiorum*, published on 20 February 1808.

13. *Study for the Barber 'Sun rising through Vapour'*, 1807
Pencil, 11.9 x 18.7 cm
Page 70 of the 'Spithead' Sketchbook
Tate Britain

From 1792 Turner travelled extensively round Britain on an almost annual basis, filling many sketchbooks with his observations. His preferred medium was graphite pencil and only rarely did he work outside in watercolour, claiming that he could make fifteen drawings in the time it took for one watercolour. Turner visited Spithead and Portsmouth in 1807 to see two captured Danish men-of-war. The present sketchbook is known to have been used in the autumn of that year as it contains a list of dates on page 2 running from 14 October to 29 November 1807. This study is the only known preliminary drawing for the Barber *Sun rising through Vapour*, which must, therefore, post-date this period. While the composition is very close to the painting it contains some significant differences of detail such as the position of the sun and of the shipping to the right.

Turner's sketchbooks served many functions and he used them casually. The majority of the pages in this one bear pen drawings of groups of ships at sea. On its back leather cover are inscribed the words 'River Thames' and 'Margate' (the latter deleted). However, the frontispiece and page 1 are inscribed with poetry by Turner and as recorded above, his use of the book as an *aide-mémoire* extended to listing the dates on which it was used.

14. *Scene on the French Coast*, c. 1806–7
Pen and ink, pencil and watercolour, 32.7 x 41.3 cm
Tate Britain

This rich monochrome watercolour bears a strong compositional similarity to the Barber *Sun rising through Vapour*, notably the sail and rigging of the boat to the left and the man at their right in the middle distance, looking out to sea. In the centre, behind the man sitting on the donkey, appears what may be the artist's thumb or finger-print. Turner often used his hands in painting his oils but less so in the case of watercolours. In the lower right margin is a second faint drawing of a port.

Scene on the French Coast is likely to have been worked up from memories of Turner's first Continental tour,

Fig. 38
Scene on the French Coast (Liber Studiorum Part I),
Birmingham Museums and Art Gallery

which took place between July and October 1802. Since he went directly to the Alps after crossing the Channel, returning afterwards to Paris, it was probably inspired by the French coastal scenes he saw just before his departure for England. Although he did not leave Britain again until 1817 he continued to produce works influenced by his visit, including beach scenes, which in turn were to inspire those of Bonington, made in the 1820s (see fig. 27, p. 27). The composition was etched in reverse with modifications by Turner and mezzotinted by his unrelated namesake, Charles Turner, for inclusion in the *Liber Studiorum* (fig. 38). It appeared in Part I of the work, published on 11 June 1807.

15. *The Drawing Room, Farnley Hall*, 1818
Bodycolour on grey paper, 31.5 x 41.2 cm
Private Collection

Farnley Hall, near Otley, Yorkshire, belonged to Walter Ramsden Fawkes (1769–1825), Turner's most loyal patron from about 1800 until his death. This view, taken in November 1818, records the drawing or music room in the new south wing, built by John Carr of York from 1786 to 1790. In pride of place over the chimneypiece is Turner's magnificent *Dort,* which Fawkes bought at the Royal Academy of 1818 (fig. 24, p. 25). Below it to the left is the Barber *Sun rising through Vapour* and opposite to the right *Portrait of the 'Victory' in Three Positions* (cat. 3). It is likely that these two works, of almost equal dimensions, were intended as pendants from an early stage.

In 1792 Fawkes, a well-connected radical politician, had inherited Farnley and about 15,000 acres. From 1806 he was Whig MP for York and already a generous supporter of contemporary artists, especially Turner. From 1803 he bought over 20 watercolours Turner had worked up from his first visit to the Alps and fourteen years later he would acquire 51 watercolours resulting from Turner's excursion to the Rhine. He eventually owned five major oils.

Turner painted thirteen views of the house and garden at Farnley, all of approximately this size, which seem to form a set, executed over the years 1815 to 1820, during which he was an annual visitor. He also worked on other projects there, notably *Fairfaxiana*, a series of watercolours commemorating the involvement of Fawkes's ancestor, General Thomas Fairfax in the Civil War (Ashmolean Museum, Oxford and elsewhere).

RELATED PAINTINGS

Julius Porcellis (c. 1609–45)
16. *Mussel Fishing*
Oil on canvas, 51 x 63.5 cm
Signed on leeboard of left boat: *IP*
National Maritime Museum, Greenwich
Acquired with the assistance of the National Art
Collections Fund

Turner was greatly inspired by the Dutch marine painters of the seventeenth century, among whom Jan Porcellis and his son Julius enjoyed a leading reputation. In the 1620s Porcellis *père* pioneered a major stylistic change from the strongly coloured works of his predecessor, Hendrick Cornelisz Vroom. Whereas he had emphasized the representation of ships *per se*, Porcellis produced monochromatic works that are essentially studies of sea, sky and atmospheric effects. In this picture his son uses a low horizon to emphasize the expanse of clouded sky behind, lacking any precise indication of the presence of the sun.

The daily routine of fishermen's lives furnished themes for many Dutch artists in the seventeenth century, while the resilience of their villages, especially in times of war, and the idiosyncratic characteristics of the Dutch coastline were symbols of great national pride. Here the anchors and mussel shells scattered on the sand – taken together with the industrious scene beneath the Dutch flags, and the shipping, town and dunes in the distance – all provide telling symbols of national security and prosperity. A number of Dutch artists painted beach scenes and fishmarkets (fig. 14) and there also existed a Flemish still-life tradition of depicting various fish and crustaceans on a beach. Turner was certainly aware of these antecedents, whose compositional and symbolic elements he frequently adapted in his own early beach scenes (see cats 2 and 6).

Willem van de Velde the Younger (1633–1707)
17. *Calm: a Dutch Flagship coming to Anchor with a*
States Yacht before a light Air, c. 1658
Oil on canvas, 87.5 x 106.5 cm
Signed lower right: *W.V.Velde*
National Maritime Museum, Greenwich

Willem van de Velde the Younger's reputation was based on seascapes of serene calmness, which dominated his early work before his move to England in 1672–3. His later works, however, are mainly storm and shipwreck scenes (see cat. 22). The present picture anticipates the tranquil atmosphere of both versions of *The Sun rising through Vapour* and the careful placing of each vessel so as to create a satisfying composition. The large ship under the cumulus cloud to the right, flying the Dutch flag on her main mast, has been identified as the *Eendracht*. In the centre foreground she is approached by a ship's boat carrying four distinguished people and a trumpeter, in honour of whom she fires a salute. In the left foreground a States yacht displays the Orange arms and behind her is a ship at anchor, believed to be Admiral de Ruyter's flagship, the *Huis te Zieten*.

In comparison with Turner, Van de Velde's horizon is low and he conceals the exact location of the sun. Juxtaposition of the small, crowded yacht and the static man-of-war to the right provides a telling compositional prototype for Turner's active group of fishermen in the Barber *Sun rising through Vapour* contrasted with the majestic warship behind them (see cat. 7). Turner saw Dutch paintings in the sale-rooms and in some of the private collections to which he had access from the 1790s. He admired Van de Velde the Younger, as did leading marine artists of the next generation such as Clarkson Stanfield.

55

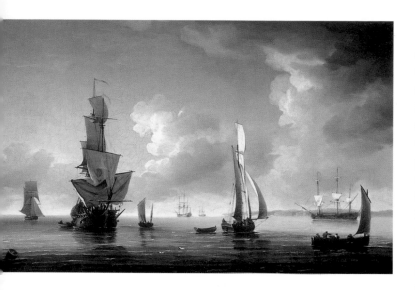

Charles Brooking (c. 1723–59)
18. *Shipping in light Airs in the Thames Estuary*
Oil on canvas, 48.5 x 78.5 cm
Signed on buoy lower left: *C. Brooking*
National Maritime Museum, Greenwich

Brooking was brought up in Deptford, not far from Greenwich Hospital and was a seaman before turning to painting. The Thames-side dockyards provided an important focus and source of subject-matter for a group of eighteenth-century marine painters that also included Peter Monamy (1689–1749) and Samuel Scott (1701/2–72). Brooking himself exhibited at the Foundling Hospital and his reputation as a marine artist was well established by 1755. His direct but imaginative response to natural effects at sea and his evidently informed knowledge of maritime practice and naval architecture led him to produce marine views at once scrupulously accurate and powerfully evocative.

This painting shows an enticing variety of sailing ships becalmed, including a hoy towing a boat in the centre and a small Indiaman bows on to the left. Unlike most of his predecessors Brooking used a light ground the better to endow his works with luminosity and atmosphere. In 1805 the watercolourist Edward Dayes wrote that 'His coloring is bright and clear, his water pellucid and transparent, with a firm, broad, spirited touch.'[1] In this respect and through his abiding interest in the Thames and its commerce, Brooking provides a link between the marine tradition established in England by Willem van de Velde the Younger and Turner. The traditional theme of shipping becalmed was adapted by Turner for *The Sun rising through Vapour*, while the buoy visible at the lower left of *Shipping in light Airs* anticipates his use of such eye-catching details in his early sea-pieces (see cat. 8).

[1] *The Works of the late Edward Dayes*, London, 1805, p. 322.

Nicholas Pocock (1740–1821)
19. *Nelson's Flagships at Anchor*, 1807
Oil on canvas, 35.5 x 54.5 cm
Signed and dated lower right: *N Pocock 1807*
National Maritime Museum, Greenwich

Pocock was brought up in Bristol and went to sea at the age of seventeen. Although he took up art only in his forties, he became an extremely successful painter of ship portraits and sea battles. This magnificent calm is close in date to the Barber *Sun rising through Vapour*, which it reflects compositionally and in the use of devices such as the sunrise in the centre and the silhouetting smoke of gunfire on the right. The serenity of the scene and the

harmony between man and nature symbolized by the pink-tinged clouds suggest a benevolent *pax Britannica* and the protection by the fleet of the smaller ships nearby. Such concerns were much in Turner's mind during the years prior to Waterloo. His fascination with the British triumph at Trafalgar and the heroic death of Admiral Lord Nelson led to the poignant *Portrait of the 'Victory' in Three Positions* (cat. 3).

Pocock's picture is a composite portrait of five of the ships on which Nelson served, showing them drying sails at Spithead, Portsmouth. On the far left with her bows visible is *Agamemnon*, which he commanded as captain. Broadside on is *Vanguard*, his flagship at the Battle of the Nile, and stern on is *Elephant*, his temporary flagship at the Battle of Copenhagen. The ship underway in the centre is *Captain*, in which Nelson first came to public attention at the Battle of Cape St Vincent in 1797. Dominating the right foreground is *Victory*, shown in her original state and not as she was at the Battle of Trafalgar (see cat. 3). With her starboard guns she fires a salute as an admiral's barge is rowed alongside. By 1807 Nelson had already been dead for over a year and the picture's elegiac atmosphere, the disparate angles of the ships' yards and the gun salute are intended to recall his passing. It is one of six painted by Pocock and engraved by James Fittler for the two-volume *Life of Nelson* published by J.S. Clarke and J. McArthur in 1809.

Sir Augustus Wall Callcott (1779–1844) after Turner
20. *Sheerness and the Isle of Sheppey*, c. 1807-8
Oil on canvas, 69.8 x 89.5 cm
Tate Britain

Turner's painting of the same title was exhibited at his gallery in 1807, whence it was bought by Thomas Lister Parker, a cousin and artistic adviser of Sir John Fleming Leicester and a friend of Walter Fawkes (fig. 39). It heralded a group of sea-pieces set in the Thames estuary, which he exhibited there in 1808, 1809 and 1810 (see cats 8 and 9). In most of these the distant town appears to be Sheerness, which stands on the northernmost point of the Isle of Sheppey at the mouth of the Medway. The present picture, together with a smaller version in the Ashmolean Museum, Oxford, was painted for Callcott's own instruction before he executed a life-size replica of Turner's original for its owner (Private Collection).[1] Callcott retains the distinctly Turnerian appearance of the fishermen, based on Flemish prototypes, together with the local colour of the fish and crabs in the bottom of the boat. His handling of paint is variable, with plenty of impasto in the manner of Turner but noticeably clumsier treatment of the bright clouds to the left. Similarly, the foaming sea in the left foreground is painted freely but the spray is more pedantic than Turner's.

By 1806 Callcott was seen as Turner's main follower and rival, though the relationship between the two was always friendly. He began as a pupil of the portraitist John Hoppner, who may have introduced him to Turner, but went on to become the most fashionable English landscape-painter of his day. Elected R.A. in 1810, in 1837 he became the first landscapist to be knighted – an honour never achieved by Turner. His early work was fresh, stressing simple composition and atmospheric, silvery lighting but his later pictures often assumed the role of acceptable imitations of Turner without his idiosyncrasies. In 1860 Ruskin was horrified to find an unofficial catalogue on sale outside the National Gallery in which Callcott's work was favourably compared with *The Sun rising through Vapour* (cat. 6).

[1] In 1848 the original was purchased by the Birmingham collector-dealer, Joseph Gillott.

Fig. 39
Sheerness and the Isle of Sheppey, Confluence of the Thames and Medway, TG 1807
National Gallery of Art, Washington, D.C.

François-Louis-Thomas Francia (1772–1839)
21. *The Old Hulk*, c. 1808
Watercolour, 30.8 x 45.4 cm
Birmingham Museums and Art Gallery

Having worked with Turner and Girtin as a copyist at Dr Thomas Monro's evening 'Academy' in the mid-1790s, Francia became a founder member and secretary of Girtin's Sketching Club. He stayed in England until 1817, continuing to exhibit at the Royal Academy until 1822. After returning to his native Calais he became a focus of contact between English and French artists, teaching Bonington for a time and maintaining contacts with David Cox, Samuel Prout and other English watercolourists.

The Old Hulk well exemplifies Francia's sombre scenes of estuaries, beaches and windswept headlands.

His subtle chiaroscuro and treatment of the water recall the style of Girtin rather than of Turner but by focusing on a single vessel a more Turneresque range of atmospheric effects in sea and sky is achieved. Abandoned warships stripped of their masts and rigging were a common sight on the coasts and larger rivers of England and feature in several works of this period by Turner, including the Barber *Sun rising through Vapour* (cat. 7). Prior to breakage for timber they were often used as floating prisons (as later described by Dickens in *Great Expectations*) or hospitals. Here a poignant contrast with the ship's former glory is reflected in the melancholy skull and detritus in the foreground, comparable with the loaded symbolism of the German Romantic landscapist, Caspar David Friedrich (1774–1840).

Elisha Kirkall (c. 1682–1742) after Van de Velde
the Younger
22. *A Ship scudding in a Gale*, 1726
Mezzotint, 63.5 x 48.3 cm
National Maritime Museum, Greenwich

Kirkall engraved a set of about sixteen mezzotints after paintings by Willem van de Velde the Younger in British collections. These were appropriately printed in sea-green ink and were in much demand during the eighteenth century. They were the way in which numerous British artists were familiarized with the paintings of Van de Velde (see also cat. 17) and thus played an important part in creating a tradition of marine art in Britain at a time when her maritime power was growing rapidly.

This is probably the print referred to by Walter Thornbury as having caught Turner's eye when he was looking over some prints with a friend. '"This", said Turner, with emotion, taking up a particular one, "made me a painter." It was a green mezzotinto, a Vandervelde – an upright; a single large vessel running before the wind, and bearing up bravely against the waves.'[1] Certainly the heaving ocean, labouring ship and agitated sky have much in common with Turner's early turbulent marines such as 'The Egremont Sea-piece' (see cat. 12).

[1] Thornbury, *op. cit.*, 1862, I, p. 18.

Daniel Lerpinière (c. 1745–85) after Claude-Joseph Vernet
23. *Calm, Sunset*, 1781
Engraving and etching, 49 x 61.8 cm
British Museum, London

In the late eighteenth century Vernet was the most important painter of the sea in Europe, and well represented in British collections. His sensitive evocations of the times of day and differing weather conditions were developed by Turner in such seascapes as the Barber *Sun rising through Vapour*, although he always avoided Vernet's idealized settings and aristocratic figures.

In 1773 Vernet sold two pictures to Lord Clive (of India). These were pendants – *Calm, Sunset* and *Storm on the Coast* – and they were engraved a few years later by Lerpinière. Published in 1781–2 by John Boydell, the prints, if not the original paintings, may well have been known to Turner. In any case, Vernet was one of the most prolifically engraved painters of eighteenth-century France, with over 300 prints after his pictures. These sold well in England, as confirmed by Boydell's remark that 'in the course of one year I imported numerous impressions of Vernet's celebrated *Storm*, so admirably engraved by Lerpinière.'[1] Although Vernet often used calm weather with a rising sun dispelling mist to represent morning, here he intends it to be evening, as confirmed by the peasant's empty basket to the left.

[1] J.T. Smith, *Nollekens and his Times*, ed. W. Whitten, II, London, 1920, pp. 184–5.

Charles Turner (1773–1857) after Turner
24. *The Shipwreck*, 1806
Hand-coloured mezzotint, 58.5 x 81.6 cm
British Museum, London

The scene was described by Thornbury, Turner's early biographer: 'A large Indiaman [is] becoming a wreck, while fishing boats endeavour to rescue the crew ... Some of the passengers are dropping from the bowsprit into the boat ... The broken rudder floats by on the dark and dirty water ... No marine painter ever painted with so sailor-like a mind as Turner.'[1] Certainly in comparison with sea-pieces of a few years before, including even *Calais Pier* (fig. 9), this dizzying maelstrom shows the rapid advance of Turner's dramatic powers. It represents the climactic moment of his early experiments with stormy seascapes, anticipating by five years the orgiastic violence and destruction of *The Wreck of a Transport Ship* (fig. 2). Its tension is much enhanced by the jagged sails and floating flotsam while the vortex that was already a favourite device of the artist here gapes open at its most terrifying, making the work the most awe-inspiring struggle of man and ocean that had yet been depicted.[2] It could hardly be further removed in mood from the shimmering calm of *The Sun rising through Vapour*.

The original painting was exhibited in Turner's Gallery in 1805 and bought at the opening of the exhibition by Sir John Leicester for 300 guineas (fig. 6 p. 12).[3] Already it had attracted the attention of the engraver Charles Turner (no relation, but a colleague from student days) who proposed publishing a large mezzotint in an edition of 50 as a speculation. It thus became the first oil painting by Turner to be engraved and was subscribed to by many of Turner's patrons and a number of artists. As a result, his reputation as a marine painter was greatly enhanced and the power and profitability of the print market demonstrated. Although this very fine impression was probably not coloured by Turner himself, he reserved this right for later ones. In 1807, when he embarked on the *Liber Studiorum*, he chose Charles Turner as his mezzotinter.

[1] Thornbury, *op. cit.*, 1862, I, pp. 266–7.
[2] T.S.R. Boase, 'Shipwrecks in English Romantic Painting', *Journal of the Warburg and Courtauld Institutes*, XXII, 1959, p. 338.
[3] In 1807 it was returned to the artist in part exchange for *The Fall of the Rhine at Schaffhausen* (Museum of Fine Arts, Boston).

William Holl the Younger (1807–71) after Turner
25. *Self-Portrait*, published 1859–61
Engraving, 16.4 x 13 cm
Tate Britain
(Reproduced as frontispiece)

Turner's only known self-portrait (Tate Britain) dates from about 1799, the year he was elected Associate of the Royal Academy at the youngest possible age. This striking engraving was issued posthumously as part of *Turner's Gallery*, a series of 61 steel engravings published by James Vertue with text by Ralph Wornum. It emphasizes the earnestness but also the burgeoning confidence of the 24-year-old artist, who uncharacteristically presented himself as something of a dandy. Painting and print brilliantly capture the moment when Turner was beginning to acquire a reputation through his marine paintings, notably his first oil shown at the Royal Academy, *Fishermen at Sea*, exhibited there in 1796 (fig. 1). Of the eleven works he showed in 1799, two were sea-pieces, though of very different kinds: *Fishermen becalmed previous to a storm – twilight*; and *Battle of the Nile at ten o'clock, when the L'Orient blew up, from the station of the gun boats between the battery and Castle of Aboukir* (both now missing). That same year Turner moved from Covent Garden to Harley Street, a more fashionable part of London and conveniently situated for prospective patrons.

James Charles Armytage (1802–97) after Turner
26. *Dutch Boats in a Gale* ('The Bridgewater Sea-piece'), published 1859–61
Engraving, 21 x 26.7 cm
Tate Britain

Turner's original oil of 1801 was one of his first major sea pictures (fig. 7). Dutch boats are shown in a fierce gale, seemingly on a collision course as fishermen try to land their catch before it is too late. The painting was commissioned by Francis Egerton, 3rd Duke of Bridgewater, as a pendant to his *Kaag close hauled in a Fresh Breeze* by Willem van de Velde the Younger (fig. 40). Turner's picture was slightly larger than its companion but assumed a similar composition, reversed for the sake of symmetry and balance. His emphasis on chiaroscuro, movement and

Fig. 40
Willem van de Velde the Younger
Kaag close hauled in a fresh Breeze, 1671–2
Toledo Museum of Art, Ohio

drama, achieved through a more aggressive painterly touch, is characteristic of his developing style. Unlike Van de Velde, Turner disposed the distant ships parallel to the picture plane, suggesting depth by the sweep of the clouds and a highly selective fall of light. He thus achieved a more structured image.

The duke paid 250 guineas for the painting (though he refused to give an extra 20 for the frame). Its admirers included the artists Constable and Fuseli, while Benjamin West, President of the Royal Academy, characterized it as 'what Rembrandt thought of but could not do.'[1] Press comment was also favourable, apart from *The Monthly Mirror*, which demanded 'greater distinctness' in the foreground boats, thus anticipating later criticism of Turner's seascapes as crude and unfinished. Canonical status for the work was confirmed in 1837 when it was lent to the Old Masters exhibition at the British Institution along with its pendant. The publication of this print likewise verified a continuing public taste for Turner's early seascapes ten years after his death.

[1] K. Garlick and A. Macintyre, eds, *The Diary of Joseph Farington*, IV, New Haven and London, 1979, p. 1539 (18 April 1801).

65

SUMMARY BIOGRAPHY TO 1810

1775: Born 23 April in Maiden Lane, Covent Garden, London.

1786: Travels begin with stays at Brentford and Margate.

1789: Admitted as a student at the Royal Academy Schools after a term's probation.

1790: Exhibits first watercolour at Royal Academy, *The Archbishop's Palace at Lambeth* (Indianapolis Museum of Art).

1791: First tour in search of topographical material, to Bristol, Bath, Malmesbury &c.

1796: Exhibits first oil painting at the Royal Academy: *Fishermen at Sea* ('The Cholmeley Sea-piece').

1799: Moves into rented lodgings at 64 Harley Street, Marylebone.

Begins liaison with Hannah Danby, who eventually bears him two daughters.

Elected Associate of the Royal Academy (November).

1800: Turner's mentally unstable mother committed to the Royal Bethlem Hospital.

1801: Exhibits *Dutch Boats in a Gale* ('The Bridgewater Sea-piece') at the Royal Academy, his first commissioned seascape.

1802: Elected Royal Academician (February).

Ships bearing up for Anchorage exhibited at the Royal Academy and bought by the Earl of Egremont; initiates patronage lasting until the Earl's death in 1837.

The Treaty of Amiens allows travel to Europe; Turner visits Switzerland and France (July–October).

1803: Appointed to the Council of the Royal Academy.

Walter Fawkes first buys Turner's watercolours; his patronage continues until his death in 1825.

1804: Buys 64, Harley Street and builds a gallery there to exhibit his paintings.

Society of Painters in Water-Colours founded to raise the status of watercolours.

Death of Turner's mother in Bethlem Hospital (His father survives until 1829).

Acquires *pied-à-terre* on the Thames, Sion Ferry House, Isleworth.

1805: Holds first exhibition in his gallery (May); it is dominated by *The Shipwreck*.

Battle of Trafalgar (21 October).

1806: Explores the River Thames in a boat, making *plein-air* oil sketches; this continues through 1807.

Two oils exhibited at the first exhibition of the British Institution.

1807: Appointed Professor of Perspective at the Royal Academy Schools.

First volume of the *Liber Studiorum* published (11 June). The fourteen parts appear irregularly until 1819.

Travels to Portsmouth to see two captured Danish men-of-war.

1811: Delivers his first course of lectures as Professor of Painting at the Royal Academy (7 January).

Commissioned by the engraver and publisher, W.B. Cooke to contribute 40 plates to *Picturesque Views of the Southern Coast of England* (issued 1814–26).

SEASCAPES EXHIBITED BY TURNER TO 1810

The Royal Academy

1796: *Fisherman at Sea* ('The Cholmeley Sea-piece'; 305); Tate Britain.

1797: *Moonlight; a study at Millbank* (136); Tate Britain (cat. 1).

Fishermen coming ashore at sunset, previous to a gale ('The Mildmay Sea-piece'; 279); location unknown.

1799: *Fishermen becalmed previous to a storm, twilight* (55); location unknown.

Battle of the Nile, at ten o'clock, when the L'Orient blew up, from the station of the gun boats, between the battery and castle of Aboukir (275); location unknown.

1801: *Dutch boats in a Gale; fishermen endeavouring to put their fish on board* ('The Bridgewater Sea-piece'; 157); Private Collection, on loan to the National Gallery, London.

1802: *Fishermen upon a Lee Shore in Squally Weather* (110); Southampton City Art Gallery (cat. 2)

Ships bearing up for Anchorage ('The Egremont Sea-piece'; 227); Petworth House, Sussex.

1803: *Calais Pier, with French Poissards preparing for sea; an English packet arriving* (146); National Gallery, London.

1804: *Boats carrying out anchors and cables to Dutch men-of-war in 1665* (183); Corcoran Art Gallery, Washington, D.C.

1807: *Sun rising through Vapour; Fishermen cleaning and selling Fish* (162); National Gallery, London (cat. 6).

1809: *Spithead: Boat's crew recovering an anchor* (22); Tate Britain.

The British Institution

1808: *The Battle of Trafalgar, as seen from the Mizen Starboard Shrouds of the 'Victory'* (359); Tate Britain

1809: *Sun rising through Vapour with Fishermen landing and cleaning their Fish* (269); National Gallery, London (cat. 6).

Turner's Gallery, 64, Harley Street

1804: *Old Margate Pier*; Private Collection.

1805: *The Shipwreck*; Tate Britain.

The Deluge; Tate Britain (?)

1806: *The Battle of Trafalgar, as seen from the Mizen Starboard Shrouds of the 'Victory'*; Tate Britain.

1807: *Sheerness and the Isle of Sheppey, with the Junction of the Thames and the Medway from the Nore*; National Gallery of Art, Washington, D.C.

The Mouth of the Thames; destroyed (?)

1808: *Purfleet and the Essex Shore as seen from Long Reach*; Private Collection, Belgium.

The Confluence of the Thames and the Medway; Petworth House, Sussex.

Sheerness as seen from the Nore ('The Loyd Sea-piece'); Richard Green, London (cat. 8).

Margate; Petworth House, Sussex.

Spithead: Boat's crew recovering an anchor; Tate Britain.

1809: *Fishing upon the Blythe Sand, Tide setting in*; Tate Britain (cat. 9).

Guardship at the Great Nore, Sheerness; Private Collection.

Fishing Boats in a Calm (The Barber *Sun rising through Vapour?*)

Fishing Boats in a Calm; location unknown.

Shoeburyness, Fishermen hailing a Whitstable hoy; National Gallery of Canada, Ottawa.

1810: *Guardship at the Great Nore, Sheerness*; Private Collection.

Hastings, Fishmarket on the Sands; Nelson-Atkins Museum of Art, Kansas City.

Sun rising through Vapour; Fishermen cleaning and selling Fish; National Gallery, London (cat. 6).

Fishing upon the Blythe Sand, Tide setting in; Tate Britain (cat. 9).

Shoeburyness, Fishermen hailing a Whitstable hoy; National Gallery of Canada, Ottawa.

Spithead: Boat's crew recovering an anchor; Tate Britain.

SELECT BIBLIOGRAPHY

Augustus Wall Callcott, exh. cat., Tate Gallery, London, 1981 (D.B. Brown).

Armstrong, Sir Walter, *Turner*, London, 1902.

Bachrach, A.G.H., 'Turner, Ruisdael and the Dutch', *Turner Studies* I, no. 1 [1981], pp. 19–30.

Bailey, Anthony, *Standing in the Sun: A Life of J.M.W. Turner*, London, 1997.

Boase, T.S.R., 'Shipwrecks in English Romantic Painting', *Journal of the Warburg and Courtauld Institutes*, XXII, 1959, pp. 332–46.

British Watercolours from Birmingham, exh. cat., Bankside Gallery, London, 1992 (S. Wildman).

Burnet, John, *Turner and his Works*, London, 1852.

Butlin, Martin and Joll, Evelyn, *The Paintings of J.M.W. Turner*, revised edition, New Haven and London, 1984.

Cave, Kathryn, ed., *The Diary of Joseph Farington*, VIII, New Haven and London, 1982.

Cordingly, David, *Marine Painting in England 1700–1900*, London, 1974.

Cordingly, David, *Ships and Seascapes*, London, 1997.

Egerton, Judy, *National Gallery Catalogues: The British School*, revised edition, London, 2000.

Finberg, Alexander J., *The Life of J.M.W. Turner, R.A.*, revised edition, London, 1961.

Finberg, Hilda F., '"Turner to Mr Dobree": Two Unrecorded Letters', *Burlington Magazine*, XCV, March 1953, pp. 98–9.

Gage, John, 'Turner and the Picturesque', *Burlington Magazine*, CVII, January 1965, pp. 16–25; February 1965, pp. 75–81.

Gage, John, ed., *The Collected Correspondence of J.M.W. Turner*, Oxford, 1980.

Gage, John, *J.M.W. Turner – A Wonderful Range of Mind*, New Haven and London, 1987.

Garlick, Kenneth and Macintyre, Angus, eds, *The Diary of Joseph Farington*, III, IV and VI, New Haven and London, 1979.

Hamilton, James, *Turner: A Life*, London, 1997.

Herrmann, Luke, 'Turner and the Sea', *Turner Studies*, I, no. 1, [1981], pp. 4–18.

Herrmann, Luke, *Turner Prints: The Engraved Work of J.M.W. Turner*, Oxford, 1990.

Hill, David, *Turner on the Thames*, New Haven and London, 1993.

Kitson, Michael, 'Turner and Claude', *Turner Studies*, II, no. 2, 1983, pp. 2–15.

Levitine, George, 'Vernet tied to a Mast in a Storm: The Evolution of an Episode of Art Historical Romantic Folklore', *Art Bulletin*, XLIX, 1967, pp. 93–100.

Presences of Nature: British Landscape 1780–1830, exh. cat., Yale Center for British Art, New Haven, 1982 (L. Hawes).

Shanes, Eric, *Turner's Rivers, Harbours and Coasts*, London, 1981.

Shanes, Eric, *Turner's Human Landscape*, London, 1990.

Taylor, James, *Marine Painting*, London, 1995.

Thornbury, Walter, *The Life and Correspondence of J.M.W. Turner, R.A.*, London, 1862; revised edition, 1877.

Turner 1775–1851, exh. cat., Tate Gallery, London, 1974–5 (M. Butlin, A. Wilton & J. Gage).

Turner, exh. cat., National Gallery of Australia, Canberra, 1996 (M. Lloyd).

Turner and the Channel, exh. cat., Tate Gallery, London, 1987 (D.B. Brown).

Turner and the Sublime, exh. cat., British Museum, London, 1980 (A. Wilton).

Turner's 'Drawing Book': The Liber Studiorum, exh. cat., London, Tate Gallery, 1996 (G. Forrester).

Turner's Holland, exh. cat., Tate Gallery, London, 1994 (A.G.H. Bachrach).

Turner: The Second Decade, 1800–1810, exh. cat., Tate Gallery, London, 1989 (R. Upstone).

Wilton, Andrew, *The Life and Work of J.M.W. Turner*, London, 1979.

Wilton, Andrew, *Turner in his Time*, London, 1987.

Young Turner: Early Work to 1800, exh. cat., Tate Gallery, London, 1989 (A. Lyles).

Ziff, Jerrold, 'Turner and Poussin', *Burlington Magazine*, CV, July 1963, pp. 315–21.

Ziff, Jerrold, "Backgrounds, Introduction of Architecture and Landscape": A Lecture by J.M.W. Turner', *Journal of the Warburg and Courtauld Institutes*, XXVI, 1963, pp. 124–47.

PHOTOGRAPHIC CREDITS

cats 1, 4, 5, 9, 10, 11, 12, 13, 14, 20, 25, 26 & figs 1, 5, 6, 13, 31, 32, 33, 37: © Tate, London 2003

cat. 2 & fig. 35: Southampton City Art Gallery/The Bridgeman Art Library

cat. 3 & figs 24, 27, 29, 36: Yale Center for British Art, Paul Mellon Collection

cat. 6 & figs 9, 19, 20: © National Gallery, London

cats 16, 17, 18, 19, 22 & figs 12, 14: © National Maritime Museum, London

cat. 21: Birmingham Museums & Art Gallery. Bequeathed by H. Watson Smith Esq., 1926

cats 23, 24 & fig. 28 © The British Museum, London

fig. 3 National Gallery of Victoria, Melbourne, Australia, Presented by the Duke of Westminster, 1888

fig. 4: In the Collection of The Corcoran Gallery of Art, William A. Clark Collection

fig. 7: Private Collector, England, on loan to the National Gallery, London. Published with the owner's permission

fig. 8: Musée du Louvre, Paris, France/The Bridgeman Art Library

fig. 10: Tokyo Fuji Art Museum

figs 11, 16, 18: V & A Picture Library

fig. 15: In a private Scottish collection. Photo: Antonia Reeve Photography

fig. 17: By permission of the Trustees of Dulwich Picture Gallery

fig. 21: Kenwood Iveagh Bequest © English Heritage.NMR

fig. 22, 26: By kind permission of the Trustees of the Wallace Collection

fig. 23: Staatliche Kunsthalle, Karlsruhe

fig. 25: National Museums Liverpool (The Walker)

fig. 34: The Nelson-Atkins Museum of Art, Kansas City, Missouri (Purchase: Nelson Trust) 31–74

fig. 38: © Board of Trustees, National Gallery of Art, Washington, D.C., Widener Collection

fig. 40: Toledo Museum of Art, Purchased with funds from the Libbey Endowment, Gift of Edward Drummond Libbey, 1977.62

All other credits are identical to the locations given in the catalogue entries.